Interdisciplinary Perspectives on Torture

At the Interface/Probing the Boundaries

VOLUME 127

The titles published in this series are listed at *brill.com/aipb*

Interdisciplinary Perspectives on Torture

Edited by

Lon Olson and Stuart Molloy

BRILL
RODOPI

LEIDEN | BOSTON

Cover illustration: Aidan Bennett. Used under Creative Commons license via Unsplash.com.

The Library of Congress Cataloging-in-Publication Data is available online at http://catalog.loc.gov
LC record available at http://lccn.loc.gov/2019907406

Typeface for the Latin, Greek, and Cyrillic scripts: "Brill". See and download: brill.com/brill-typeface.

ISSN 1570-7113
ISBN 978-90-04-40570-7 (paperback)
ISBN 978-90-04-40801-2 (e-book)

This book is printed on acid-free paper and produced in a sustainable manner.

Contents

PART 3
Historical and Literary Perspectives

Acknowledgements

The editors extend their appreciation to all of the contributors. Special thanks go to Glenn Ayers and Diana Medlicott. Glenn did much of the early work to prepare the manuscript for publication. Diana not only contributed an outstanding chapter to the volume, she also convened the conference on torture sponsored by Interdisciplinary.Net and held at Mansfield College, Oxford in July 2016. The papers presented form the basis of this collection. Stuart would also like to thank The University of Western Australia for the support provided in funding both his research and his attendance at the conference. Finally, the editors express their sincere appreciation to the team at Brill for their endless patience working through numerous editorial challenges.

Notes on Contributors

W. Fitzhugh Brundage
is William B. Umstead Professor of History at the University of North Carolina, Chapel Hill. He has written on lynching, utopian socialism, and historical memory in the modern American South. He is currently completing a history of debates about torture in the United States from the age of contact to the 'War on Terror.'

Federico Ciavattone
is an independent scholar affiliated with the Italian Society for Military History. His research interests are centered on military history, modern Italian and European history and the history of gender, particularly the role of women in the Second World War.

Toni Koivulahti
is a doctoral candidate at the Department of Systematic Theology at the University of Helsinki. His research interests include varieties of postmodern political theology and the interconnections of religion and violence.

Diana Medlicott
is an independent scholar working in London, UK.

Stuart Molloy
is a doctoral candidate in English & Literary Studies at The University of Western Australia. He is studying the representation of psychopathic violence in Anthony Burgess' *A Clockwork Orange* (1962), *American Psycho* (1991) by Bret Easton Ellis, and Bryan Fuller's TV series *Hannibal.*

Lon Olson
earned his PhD from the University of Texas at Austin and has taught ethics and philosophy at the U.S. Naval Academy. He is currently the Assessment Director at Texas State University and lectures in the philosophy department. A former naval officer, his research interests include virtue theory, war crimes, moral injury and just war theory.

Martin Previšić
is an assistant professor in the Department of History at the Faculty of Humanities and Social Sciences of the University of Zagreb. His current research

interests are focused on the history of the Socialist Yugoslavia and the International Communist movement.

David Senesh
is a certified clinical psychologist and supervisor of psychotherapy in Israel. He is certified by the IRCT to apply and train professionals in the application of the Istanbul Protocol. He is based at the Levinsky College of Education in Tel Aviv and the Jewish Theological Seminary in New York City. Dr. Senesh is also a survivor of torture as a former Israeli prisoner of war in Egypt in 1973.

Noora Koivulahti
is a doctoral candidate in social ethics at the Department of Systematic Theology at the University of Helsinki. She is currently working on her dissertation, focused on the comparability of animal and human rights.

Hedi Viterbo
is Lecturer in Law at the University of Essex. His research interests include state violence, childhood, and sexuality.

Introduction

It is a sad commentary that two decades into the 21st Century we are writing an introduction to yet another volume that examines the malign influence of torture on human society. Many who strive for the realization of humanist values nurtured the hope that the employment of torture as a weapon of state coercion had been relegated to a handful of rogue authoritarian regimes. At the same time, the fact that torture has proven so stubbornly resistant to eradication should come as no surprise. Torture has been, and always will be, an unfathomable feature of the human psyche. Evidence of torture's prevalence predates recorded history and in the era of cable and internet news we are confronted daily with the stark evidence that the practice of torture continues to thrive, particularly in war-torn areas where the rule of law has been forsaken. Nonetheless, many of us believed that those countries founded upon democratic principles of government were reliably inoculated against the institutional practice of torture and, thus, formed a vanguard in the ongoing campaign against its practice. Unfortunately, this optimism dimmed in the months following the attacks of September 11, 2001, when the United States deliberately embarked on a policy that legitimatized the torture of detainees. By sophistically manipulating the legal definition of torture ratified in international statutes, attorneys for the U.S. Justice Department's Office of Legal Counsel (OLC) approved a range of highly coercive interrogation practices that were specifically prohibited by international law. In one of the most regrettable chapters in U.S. history, American policy makers, and those charged with executing that policy, endorsed and engaged in brutally coercive interrogation practices that are morally beyond the pale of civilized society. The alacrity with which American policy makers resorted to torture demonstrated that, far from being eradicated, the concept of torture as a morally acceptable method of interrogation remained viable, even in one of the world's oldest democracies. The U.S. policy of employing torture was eventually rescinded. However, there is still support for its use in many quarters of the U.S. national security apparatus as well among many members of the American public. In short, the threat of torture's resurgence as a tool of state policy is far from over.

Against this backdrop, and motivated by a desire to better understand and mitigate the ongoing threat posed by the practice of torture, an interdisciplinary conference was assembled at Mansfield College, Oxford in the summer of 2016. Some two dozen scholars, representing diverse national backgrounds and academic disciplines, met to present and discuss papers that examined the practice of torture through literary, legal, philosophical, theological and

 | DOI:10.1163/9789004408012_002

historical perspectives. Ten of the papers presented at Mansfield, selected to reflect the interdisciplinary discussion that ensued, comprise the chapters of this book. The first chapter of the volume begins with an examination of torture from the criminological/legal perspective. In "Death Row Inmates: Victims of Torture in the Land of the Free," Diana Medlicott argues that the processes concomitant with the administration of the death penalty, particularly as practiced in the United States with its solitary confinement, decades-long appeals process and frequent and arbitrary stays of execution, results in psychological effects on prisoners that are a form of systematic torture condoned by the State. In Chapter Two, "Torture's In/Visibility", Hedi Viterbo examines the contribution of real-time photographic and video images to the formation society's perceptions of the practice of torture and how privileging real-time photographic images (which are often times scarce) in courts of law over alternate visual methods of documentation hampers the prosecution of torture as well as draws public attention from the social, political and legal structures that enable the practice of state torture to continue.

Chapters Three through Five bridge the space between legal and philosophical/theological perspectives. In "The Ambiguity of Sovereignty: The Passion Narrative as Paradox," Toni Koivulahti analyzes the Christian passion narrative, with its description of the torture and crucifixion of Christ, through Jacques Derrida's conceptual framework of the three figures that exist outside of the law: the beast, the criminal and the sovereign. He argues that viewing torture through this lens opens up new narratives regarding the universal experience of torture and aids in understanding how ritual violence is transformed into actual physical violence. In Chapter Four, "Barbarians at the Gate: Reasserting a Natural Law Definition of Torture," Lon Olson argues that the responsibility for the moral violations committed by U.S. interrogators after September 11, 2001 must be shared with the lawyers for the U.S. Justice Department who crafted an excessively restrictive definition of the degree of pain required for an interrogation technique to qualify as torture, thus permitting the employment of a plethora of coercive techniques previously prohibited. In contrast, Olson argues that had the attorneys adhered to natural law principles that promote the inherent dignity of human life and reason, a great deal of unnecessary suffering would have been prevented. Chapter Five rounds out the philosophical discussion with Noora Virjamo's "Beyond the State: Human and Animal Positions Outside the Law" in which she describes how the practice of state torture is initially enabled by placing the intended victims outside the circle of legal protections accorded to citizens. She reinforces this point by comparing the practice of placing ethnic groups that have historically been the victims of extra-legal, discrimination, persecution and torture, to the plight of animals in modern society, who similarly lack formal, standing legal rights, and are subject to institutionalized torture through animal experimentation and agriculture.

Chapters Six through Eight approach torture from a historical perspective, with an eye to how the practice of torture has shaped modern society. In "Torture, Slavery, Civilization and Human Rights in the United States 120–1860", W. Fitzhugh Brundage examines how torture was inseparable from the practice of slavery in the antebellum South. To illustrate this, Brundage uses a horrific example from New Orleans in 1834, and how abolitionists capitalized on the case, and others like it, to argue against the view of slavery put forth by its proponents as a benign, paternalistic institution that was necessary to ensure the slaves' wellbeing. Brundage argues that exposing the symbiotic relationship between torture and slavery resulted in increased public revulsion with the practice of slavery, especially among many northerners who had remained ambivalent towards the practice and contributed to the eventual abolition of the practice and the advancement of a broader conception of human rights. In Chapter Seven, "Torture and Anti-Partisan War: The Case of the Italian Social Republic 1943–45," historian Federico Ciavattone, describes how torture was an integral feature of the civil war that unfolded between the Italian Socialist Republic (ISR), founded by Benito Mussolini, and the monarchical Southern Kingdom established by 'Victor Emmanuel III' and 'Pietro Badoglio'. Conducted in parallel with the Second World War, this internecine conflict saw the widespread use of torture by the ISR to eradicate political opposition and quell partisan attacks against the ISR and the German combat units that were instrumental in propping up the ISR's rule in northern Italy. In Chapter Eight, "The Goli Otok Camp: Torture Justified by External Threats", Martin Previsic describes the reeducation camps established by Tito after his ideological split with Stalin in 1948, in order to root out Stalinist sympathizers that posed a threat to Tito's regime. Ironically, a key component of these anti-Stalinist reeducation camps was the prominent reliance on Stalinist techniques of abuse and torture designed to impose political orthodoxy. The existence of these camps is a continued subject of discussion to this day and, along with America's use of torture after 9/11, illustrates the ease with which torture is justified when society is faced with perceived existential threat.

In the final chapter, "A Real Show of Horrors: Reading Representations of Torture in *A Clockwork Orange* and American Psycho," Stuart Molloy examines the portrayal of torture in two popular novels and compares their depictions of torture with three diverse, but canonical sources, with each of these sources articulating three essential requirements that acts of abuse generally must satisfy in order to be considered torture. What emerges from this analysis is the central role of imagination in the practice of torture, particularly its genesis in the imagination of the torturer and the power that imagination exerts in exacerbating the psychological suffering of the torture victim. Given that the motives and means of torture first arise in the imagination, and considering the imagination's almost limitless scope, it is unrealistic to hope that the practice of torture will ever be totally abolished, only managed and mitigated.

An interdisciplinary approach to the study of torture is valuable for a few reasons. Attempting to understand torture from only one perspective is akin to examining an organism through a microscope: the microscope magnifies one particular aspect of the organism, while at the same time obscuring other salient features. In a similar fashion, critiquing the practice of torture as a narrowly legal issue, as demonstrated by the OLC, risks overlooking important features of the practice that transcend definition by legal norms and terms.

As noted above, torture has always been part of the human experience and the bones of the dead bear mute testimony to our proclivity to torment our fellow humans. During a 2005 excavation of a 7,000 year old mass grave in Germany, archaeologists found clear evidence of brutal, systematic torture among the skeletal remains of thousands of people. The practice of torture is chiseled in inscriptions on Mesopotamian monuments, and manuscripts indicate that torture was a sanctioned tool of interrogation in the ancient Sumerian legal code, *Ur Nammu.* In fact, many ancient and medieval courts of law viewed confessions extracted through torture as more reliable than those given voluntarily; hence, they were given greater weight in judicial deliberations of guilt and innocence. Torture is endorsed in passages of many of the world's sacred texts and has been leveraged as a tool for combatting religious (and political) heresy. Torture was sadistically employed by priests-interrogators during the Spanish Inquisition, where the *auto de fe* (act of faith) was incorporated into the religious life of the community as part of a 'reconciliation' process that endeavored to turn the torture of the condemned into a cathartic release from the sin of heresy, both for the accused and for the public that witnessed the torment. Politically, meanwhile, torture continues to be the preferred method of instilling the terror necessary to enforce orthodoxy within totalitarian regimes.

Torture has been the topic of numerous works of fiction that have become part of the literary canon. Edgar Alan Poe's short story, *The Pit and the Pendulum,* provides an example. Loosely inspired by the Spanish Inquisition, the story painstakingly documents the psychological torment the protagonist experiences and the exegesis of Poe's tale of horror has been a rite of a passage for generations of high school English students. More recently, the American television series, *24* did much to establish the belief among many members of the viewing public - which includes policy officials! - that torture is an efficacious means for government security forces to extract information from recalcitrant prisoners.

These examples, and others like them, have not only influenced our conception of torture, but have also comprised the context for the development of the legal codes, historical identity, modes of literary expression and the theological and philosophical concepts that underpin our sense of ourselves, our society and what it means to be truly human. Our views of torture are the consequence of thousands of years of human history and coming to an accurate understanding of the inimical nature of torture, and the genesis of current laws established to prohibit it, requires an ongoing examination of the practice from all of the humanistic perspectives represented above. While these disciplines approach the subject of torture from different vantage points, they arrive at the same conclusion: torture represents a frontal assault on the human spirit and, by extension, an assault on the health and integrity of human society. To ignore the lessons imparted by the humanities on the dangers of torture is to run the risk of sabotaging the social progress of generations and relapsing into a darker period of human history.

Regrettably, current attitudes entail that very risk becoming reality as public apppreciation for what the humanities contribute to society is waning. This decline has several sources. The cost of a college education has risen dramatically over the course of the last thirty years and increasing numbers of students are drawn to science and technology, out of sincere interest no doubt, but also driven by the reasoning that these fields provide concrete, marketable job skills that are attractive to future employers and offer a more reliable return on their investment in education. A consequence is that many view the study of the humanities as an unaffordable luxury, if not a wasteful extravagance. This view is reinforced by legislators at the local and national levels, who control funding for public universities and view an education in the humanities as either an inefficient use of taxpayer money or a form of liberal indoctrination. College administrators, caught between coping with slashed appropriations for public universities, while simultaneously trying to satisfy the demand signal from a new generation of students, find themselves aiding the demise of the humanities. One sign of this decline is that many smaller colleges are no longer replacing tenured humanities faculty when they retire. Others are doing away with entire major courses of study in the humanities along with the departments that support them.

There is no clearly discernable trail linking the decline in humanities education and the reemergence of torture as a morally acceptable public policy measure. However, it is worrisome, as well as oddly coincidental, that we find ourselves confronting the same tired arguments for torture that were thought to have been dealt with and dispatched generations earlier, all at a time when an appreciation for the very set of disciplines devoted to promoting human worth and value are on the decline. Despite the lack of a clear causal link, it seems that one sure path towards ensuring that the concept of torture reestablishes a firm toehold in the modern moral consciousness is to continue to do nothing to stem the demise of a set of disciplines that has perhaps done the most to document torture's inimical effects on human society and to argue against its continued acceptance and practice.

Torture is just one of many threats that confront humanity. We have been living with the threat of nuclear war for the last seventy-five years. Over the last three decades, the consequences of human-induced climate change have become increasingly felt. We are currently in the midst of a mass extinction of species, whose role in maintaining the health of the planet we do not fully understand. In addition to these challenges, exponential advances in the development of biomedical gene manipulation and cloning, combined with parallel developments in artificial intelligence, are forcing us to confront what it means to be human and to ask unsettling questions about what we are willing to risk in order to preserve that humanity, as imperfect as it may be. While science and technology are indispensable to developing solutions to many of these problems, without an understanding of the human condition that the study of the humanities provides, we will be navigating such dangerous waters with an inadequate compass.

Providing a strategy for a renaissance in the humanities is far beyond the scope of this book. Nonetheless, one way to reinvigorate the humanities is to continue to tackle our toughest moral and existential challenges and demonstrate what an interdisciplinary, humanistic approach has to offer. These problems all need to be attacked simultaneously, and torture is just one battle in a larger war. In this effort, the editors and contributors hope that this volume adds valuable insights to the continued campaign to make torture something we read about in history books, rather than confront in the news.

Lon Olson
Texas State University
January 2019

PART 1

Criminological, Legal and Psychological Perspectives

CHAPTER 1

Death Row Inmates: Victims of Torture in the Land of the Free

Diana Medlicott

Abstract

This chapter stems from reading accounts by prisoners on Death Row in the United States, which shows the processes and machinations that contribute to the prolonged and agonising nature of life on Death Row, and the psychological effects produced by brutal regimes and processes. It is argued that the only way to understand life on Death Row is as a form of torture operated, regulated and controlled by the State. Those that are targeted are distinctive in terms of race, poverty, deprivation, marginalisation and lack of access to proper legal defence. In addition, it is claimed that the death penalty, as practiced in the United States, fails the usual tests for the institution of punishment, and retention of it is counter-rational. Nevertheless, it has become domesticated and made ordinary in those states where it is practised, a visible and accepted cultural artefact, albeit in a fantasised version. Its true horror remains unseen, invisible and misunderstood. The reasons for its survival, which run so counter to the trends in Europe and in other industrialised nations, are suggested as originating partly in historical processes of state formation, partly in particular cultural discourses surrounding race and crime, and partly in features of localism and the politics of fear. Bringing proper visibility to this phenomenon is an important task, made possible by the voices of those on Death Row.

Keywords

prisoners – death row – United States – torture – justifications – retention – counter-rational – domesticated – state formation – cultural discourses – localism – visibility – voices

 | DOI:10.1163/9789004408012_003

Death Row

Of the 139 countries worldwide, two thirds have now abolished the death penalty. The United States reintroduced the death penalty in 1976, and since that date, 1,476 people have been executed. As of July 1st, 2017, there are 31 death penalty states. In addition, the U.S. military and the U.S. federal government can both legally impose death sentences. While the U.S. military rarely imposes the death penalty, Texas has executed more prisoners since 1976 than the next six U.S. states combined.[1] There are about 25,000 murders a year in the U.S. and about 250 to 300 death sentences delivered, which are often the product of highly dubious and arbitrary decision making. The most significant factor will be the race of the defendant and of the victim. The majority of people on Death Row are in southern states, such as Texas and Florida. In terms of ethnicity, 41% of Death Row inmates are Black, 13% Hispanic, and 42% White.

The Death Row population numbers over 3500 at any one time, with about 50 being female; however, executions are comparatively rare. About one third to one half of those on Death Row have their sentences commuted to life without parole, and a very few have their convictions overturned. Since 1976, 152 people have been exonerated with evidence of innocence, usually because of intense legal attention to the cases, such as the Capital Appeals Project. Most inmates cannot afford their own attorney and have to do all the appeals themselves. The appeals process is tortuous, and takes years to complete, and legal representation during the appeals process is often inferior. The process has nine tiers, with three of the stages capable of being gone through three times. At the very last stages, when a prisoner has received an execution date, there can be stays of execution, of 24 or 48 hours. One prisoner received four stays in five days, and was eventually executed after this torture.[2] The following extract from a letter gives some idea of the suffering these delays cause:

> I occasionally go through these bouts of confusion, fatigue, terminal nervousness and inability to focus a single thought. Serious or not, I'm afraid I'm letting these execution dates I've got hanging over my head get to me. Like so many people down here, sometimes I think I'd rather have it carried out and over with than keep facing it again and again and again.[3]

---

1 Figures relating to the death penalty in this chapter are, unless otherwise stated, drawn from www.deathpenaltyinfo.org, accessed 22nd June 2018.

2 Jan Arriens, *Welcome to Hell: Letters and Writings from Death Row* (Boston, MA: Northeastern University Press, 2015), 164. This suffering was endured by Raymond Clark in Florida State Prison, and it is one of several similar harrowing cases in Arriens 2005:164.

3 Ibid.

The rarity of actual executions is not the point: it is the operation of Death Row, with its arbitrary ambivalence, its studied uncertainties, its mismanagement, and its bureaucratisation that, taken in combination, is a form of systematised torture approved by the State, despite the Eighth amendment to the United States Constitution ruling out cruel and unusual punishment.[4] Many inmates in their thirties have spent their entire lives on Death Row, having been arrested for crimes they allegedly committed as juveniles. The Death Row population is largely composed of paupers, whose childhoods were filled with intense pain, abuse and humiliation. Many have no families or friends and are completely cut off from any kind of life. Every day for ten years or more, each prisoner wakes up knowing that the State is striving with every means in its power to take his or her life. It is this knowledge, and the uncertainty of the appeals process that constitute torture, in combination with the terrible conditions and relentless brutality that exist on Death Row.[5] A prisoner describes enforced hair cutting on Death Row in Mississippi:

> Officers and guards are going around taking people out of their cells, cutting their hair for no good reason. The ones who don't want a haircut, the guards beat them down with four-foot sticks after handcuffing and leg-shackling them. Then they cut their hair anyway.[6]

Why Is the Death Penalty Retained?

In penology, the application of punishment is considered to need justification. This is because to do intended harm to individuals in a civilised society requires moral underpinning. The standard justifications for punishment are deterrence, retribution and rehabilitation. We can examine the deterrent effect quite easily by comparing states with the death penalty and states without it. The majority of death penalty states show murder rates higher than non-death penalty states. The average of murder rates per 100,000 population in 1999 among death penalty states was 5.5, whereas the average of murder rates among non-death penalty states was only 3.6. States without the death penalty fared better over a whole decade. Nevertheless, in the past ten years, the number of executions in the U.S. has increased while the murder rate has declined.

---

4 U.S. Code, 18 USC 2340, http://www.law.cornell.edu/uscode/uscprint.html.

5 This summary of conditions is drawn from first-hand accounts in Arriens 2005.

6 Willie Reddix in Arriens 2005:49.

The death penalty, as it is practised in the U.S., is not purely retributive because of the delays between the crime and its penalty, which can be more than ten, twenty or thirty years. In these years, there is no closure for victims' families, so retribution is attenuated when they would wish it to be absolute. The delays bring great suffering to them. Clearly, it is also not rehabilitative. So, it fails the standard justifications for the institution of punishment. What we can say is that it is extremely expensive. In California, the system costs tax payers 114 million dollars a year. Taxpayers have paid more than 250 million dollars for each of the executions so far in that state. In Texas, a death penalty case costs an average of 2.3 million dollars, which is three times the cost of imprisoning someone in a single cell for 40 years. However, we could view the death penalty as an assertion of state or sovereign authority. But this is not a convincing argument, for the application of the death penalty is extremely haphazard and ambivalent. From only three thousand people, only about one hundred are executed each year, so it is an ambiguous expression of authority, and not a confident or public demonstration. Indeed, in its very privacy, it seems designed to placate the victim's family, as they are amongst the few allowed to witness it.

Neither can support for the death penalty be argued as a case of historical continuity, although the racialisation involved suggests some continuity with lynching. The death penalty is commonest in the American South, with an overwhelming over-representation of black Americans sitting on Death Row. But it is hard to argue that it is a case of historical continuity, because it is so very different from lynching, which was a form of rough and ready vigilante "justice", witnessed by the community. By contrast, Death Row is rigorously regulated by procedures and rules, and it is a private procedure. It is not done at the whim of a community, like lynching, but rather a long-drawn-out ordeal, with a lengthy appeals process. Unfortunately, this enables defenders of the death penalty to argue that, because it is hedged about with so many bureaucratic rules, and because it is increasingly carried out by means of a lethal injection, it is a civilised penalty.

So why does the United States, this advanced capitalist state, still maintain this systematised form of torture after all of Europe and most of the rest of the world have abandoned it? And, in states where it is retained, why is it so popular? Why are politicians so afraid as to risk their credibility by standing out against it?

For some sort of explanation, we can look at Norbert Elias's work, *The Civilising Process*, which was the first formal analysis and theory of civilisation. In the second volume, he looks at habitus – that package of manners, behaviour, everyday mores, which have changed so markedly over the centuries. For instance, violence has sharply decreased over the centuries, and the way we conduct ourselves in the company of others has also changed. Manners have increasingly become more refined, and interestingly we often use the word civilised when we just mean well-mannered.[7] David Garland uses Elias's work to explain the death penalty in a symbiotic relationship with state formation and distinguishes three eras.[8] The first is the Early Modern period, which means the 18th century. In this period, we see absolutist leaders using executions and torture to affirm the power of the emerging nation, and the increasingly centralised nature of the state. Thus, the emerging state tries to establish a monopoly on violence. A good example is the torture and execution of Damiens, who attempted regicide, at the start of Foucault's *Discipline and Punish*, which is impossible to forget once read. It was not enough to torture Damiens with boiling oil and pincers, it was not enough to disembowel and draw and quarter him – his remains had to be burnt to dust in order to demonstrate the power of the State and Crown.[9] During the second period, from the late 18th century till the late 20th century, we see states reducing their use of the death penalty. The state-building process has been successful, the hegemonic power of elites has been established, and with the resultant pacification of the population, more humane criminal justice institutions are developed (such as the prison). In this period, capital punishment is a performative crime-control sanction, meant to deter and be retributive.

In the third period, the last decades of the 20th century, most European countries abolished capital punishment. Until mid-way through this period, the U.S. seemed in line with other western nations. Indeed, some U.S. states had been in the vanguard of abolition – with Michigan being the first English speaking state in the world to abolish it, in 1846. In 1972, the U.S. Supreme Court declared the application of the death penalty to be unconstitutional on the grounds that it was a cruel and unusual punishment. But by the late 1970s, with a backlash against the civil rights' movement, and crime starting to rise again, many American states managed to overcome this constitutional hurdle and re-establish the death penalty, and thus they keep company with Iran, Saudi Arabia, Pakistan, and China, who, with the United States, are responsible for 93% of the world's executions.

So why did some states in the U.S. diverge from the path of other developed industrial societies? We can distinguish a peculiar process of central state

7 Norbert Elias, (*The Civilising Process, Vol. II, State Formation and Civilization,* Oxford: Blackwell 1982), 303.

8 David Garland, *Peculiar Institution: America's Death Penalty in an Age of Abolition,* (Cambridge, MA: The Belknap Press of Harvard University, 2010), 70–100.

9 Michel Foucault, *Discipline and Punish,* (Harmondsworth: Penguin, 1973), 2–25.

formation which sets the U.S. apart from other industrial and democratic societies. While no longer the case, historically, the U.S. federal government has been comparatively weak, with the individual states exercising embedded institutionalised power at the local level, limiting the powers of the central state. From the perspective of Elias, in terms of manners and ways of behaving, the U.S. does not fit the usual model, having had no monarchical or court elite. As a result, there has been no centralised nobility acting as a model of manners and behaviour. Thus, localism, with all its extremist values, has flourished. Localism has meant that traditions of popular or communal justice have survived, and local politicians have built on this, wielding powers that do not exist regionally in European states. District attorneys have been able to wield unusual powers, and demagogic local politicians have been able to build on local populism, often exploiting racial divisions and celebrating executions as an expression of localised democracy. Often, they are claimed to be necessary to save populations from dangerous criminals, who are typically portrayed as African-Americans. A backlash against the civil rights movement helped to reinforce this process, as did a carefully fostered anti-elitism. This combination of factors has fostered other, emergent strands of extremist thinking that include subjects as diverse as science versus creationism, abortion, and gun control. Thus, a clever mix of reactionary populist politics, racism, a desire for strong local authority and an aversion to central government have made the death penalty a useful electoral tool in local state politics. In turn, this has often encouraged demagogic politicians to build careers using fractured race relations and the politics of fear.

Domestification

The claim that Death Row is a state-sponsored system of torture is not really a contentious one. Indeed, it is blindingly obvious. The deficiencies of this American criminal justice institution are so often evidenced in many academic fields that it is no longer news. But that the system continues, in the face of insurmountable evidence indicating its short-comings, is a matter of shame, and, indeed of curiosity. Why, with its expensive bureaucracy, its lumbering injustices, its discriminatory practices, does it persist?

For it to persist, it must be spoken of in mundane fashion, talked about in everyday terms, rendered familiar, and thus acceptable. This very visibility, however, actually renders the reality of Death Row invisible. How, then, has this dual process of (in)visibility happened, so that this awful institution has ostensibly become so visible, ordinary and routine, whilst its awful reality has been effectively rendered invisible?

We have considered the connections with centralised State formation, but just as interesting is the way in which the death penalty has been domesticated, and here we should consider America's propensity for story-telling. All societies tell stories about who they are, and this is how, in a Hegelian sense, a community makes itself, renews itself, incorporates change and produces a narrative of continuity. But, in the U.S.A, the story-telling and the dominance of film-making are peculiarly striking, in that they seem to depict a fantasised image of American society.

Stephen Mennell is a writer who has focused on America's misperception of itself.[10] He points out that billions of people outside the U.S. know an enormous amount about America, its Constitution, its politics, manners, and its culture, so visible to all. But it is a one-way mirror: Americans cannot see the outsiders watching them. They are caught up in their own brightly lit reflection and are emotionally very satisfied by that. To support this claim, survey data shows that a large proportion of Americans do not see out very clearly and hold quite crude, stereotypical and Manichean views of the outside world. Even well-educated Americans have difficulties in mastering the geography of Europe and the rest of the world. It can, of course, be argued that a quintessential super-power has no need to know much about the rest of the world. Mennell quotes Karl Deutsch's declaration that power means the ability to talk instead of to listen, and power means being able to afford not to learn. But the drawback of an impervious stance such as this is that a nation is unable to recognise internal weaknesses. This propensity for story-telling, together with other cultural and political factors, has produced a fantasised account of crime, criminals, race, incarceration and execution, and a huge ignorance of what is entailed by such a fictional account. Movies, the popular press and other cultural artefacts are powerful tools in delivering emotional education, and if they consistently fail to portray reality, then opportunities are lost to instil empathy and understanding of human rights. Fantasised visions of crime and punishment produce a thirst for vengeance as an easy answer to difficult questions. Reality, in these visions, is sacrificed. The terrible standards of legal representation, the madness produced by life on death row, the shocking failures in human rights – these are not themes which are accurately tackled by populist ideology or media representation. Nor is there a widespread recognition of what working on Death Row does to corrections officers. The average life expectancy of this occupational group as a whole is 59, and stress,

10 Stephen Mennell, 'An Exceptional Civilising Process' *Journal of Classical Sociology* 2009, Vol 9: 1.

hypertension, alcoholism and suicide are common. It is reasonable to suppose that those who have to work on Death Row are in an even worse occupational environment than corrections officers who work in prison.

Narrative

In the past, I have written at length about the ways in which those trapped in the criminal justice system are made the object of official narratives.[11] These narratives are often legal, but also medical, judicial and psychiatric, and those who are objectified may find these accounts radically alien to their life experiences, and their subjective view of themselves.[12] I propose now to look at three significant stages of narrative for those who have arrived on Death Row, and briefly point out how, at each stage, personal autonomy has been cast and re-cast in different ways. In order to do this, I must necessarily compress and abbreviate all the complex stages of crime, trial, verdict, sentence and aftermath, and pick out those elements which demonstrate how personal autonomy becomes an elusive entity, first of all emphasised to bolster the prosecution's case, then diminished to argue for mercy, and perhaps only achieving an ironical power when speaking from Death Row, where, ostensibly, all everyday autonomy is forcibly removed. In this compression of a long and complex process, I draw on Austin Sarat's extraordinary book, *When The State Kills*. That we can hear these powerful voices is a kind of miracle, given how closed off from normal society Death Row remains. What they have to say confirms the contention of this chapter – that Death Row constitutes a slow form of torture, often drawn out over many years.

The first stage of narrative is the story constructed by the prosecution within the trial itself. This is usually a linear narrative in which the accused is seen as possessing a malign autonomy, and armed with that malignity, takes a series of steps that lead inexorably and intentionally to a fatal crime. The crime is related as a story of evil actions carried out by an evil person, but one who is nevertheless a reasoning and autonomous individual, one who could so easily have chosen the path of virtue and acted otherwise. This is a narrative that is based on populist understanding and common-sense thinking. It is the stuff of countless movies and other cultural artefacts, in a country which, as I suggested earlier, is proud of its story-telling, in thrall to its movie industry, and addicted

---

11 Diana Medlicott, *This Thing of Darkness: Perspectives on Evil and Human Wickedness*, (New York: Rodopi 2001), 85–90.

12 Diana Medlicott, *Surviving the Prison Place* (Aldershot: Ashgate, 2001), 45.

to fantasised images of good and evil. Populist narratives authorise the application of the severest punishment as the only just response. The moral context is provided by the grieving family of the victim, and this endorses the simple requirement of absolute retribution – an eye for an eye, a life for a life.[13]

Once the guilty verdict has been reached, the capital trial enters the penalty phase. At the heart of this stage is a small group of lawyers who devote their professional lives to saving from execution those who have been put through trial and found guilty. Now that narrative of malign autonomy is opposed by a somewhat feebler one – feebler, ironically, just because it is more complex and attenuated. This narrative must try to gain a foothold in a culture whose dominant ideology values simplistic stories of right and wrong, and in a specific setting where the inevitable response to crime must be the harshest punishment. Death penalty lawyers must attempt to breathe humanity into some of the most hated people in American society. Ironically, they can only do this by removing some autonomy from the guilty and recasting them as individuals who have endured a life of violence, deprivation and pain.

This defensive narrative speaks of the social conditions that have brought about the crime and conviction, and in so doing will attempt to water down the emphasis on malign autonomy, and instead present an account of severely compromised autonomy. Instead of being a person who has freely chosen evil, the convicted may be presented as someone almost unable to make choices because of the damage inflicted throughout a life of pain and hardship. Some or all of the following will be delineated as significant features in the life of the convicted – an impoverished and abusive childhood, racial prejudice, chaotic or absent parenting, learning difficulties, mental health issues and the perils of addiction. Evidence of these features will be presented to the court.

Of course, this narrative of one individual life is also re-telling, in a sense, a national narrative. It speaks of America's most glaring injustices. This national narrative, embodying widespread institutional racism, huge differences in social and economic life chances, and significant discrepancies in physical and mental healthcare, is usually particularised in the lives of individuals convicted of terrible crimes. But it is also a national narrative that challenges America's idealised view of itself as a fair and just society. So, the particularised narrative of a convicted criminal is an unwelcome story. It is an uncomfortable departure from the idealised view of America and its Constitution. It asks if the law can afford a degree of mercy or protection to those most despised

---

13 Austin Sarat, *When the State Kills: Capital Punishment and the American Condition*, (Princeton: Princeton University Press, 2001), 93.

members of society, who may have committed unforgiveable crimes but who still cry out for mercy. It asks for a sentence that falls short of a death sentence. Behind the courtroom drama, then, lies the question: what is most satisfying to the national psyche – mercy or vengeance?

The subtlety of the narrative of compromised autonomy is a difficult one to navigate. No one mitigating factor can be claimed as entirely causal, although in combination it can be powerfully argued that these mitigating factors have severely limited the autonomous capacity of the accused. But these arguments, and the evidence for them, are complex and specialised, grounded in scientific rather than populist thought. The differences between correlated factors and causative factors can be said to constitute a problematic at the very heart of the social sciences, and it is not a problematic which translates easily into a plea for mercy. Nor can it stand up to the strong simplicity of an unambiguous narrative, where an evil person killed an innocent victim.

At the same time, as presenting a narrative of autonomy that has been severely comprised by adverse life experiences, death penalty lawyers must demonstrate a necessary acceptance of the crime and the guilt of their client. For, whatever the quality of the defence representation was at trial (which, according to Clive Stafford Smith in the Appendix of *Welcome to Hell*, may have included representation by law students, lawyers who nod off to sleep or drink alcohol during the trial, and those lawyers who are incapable of any other form of gainful employment), by this sentencing stage, the crime has been "proven" to have been committed by the accused. So, the terrible violence of a crime which caused loss of life must be meaningfully recognised by the death penalty lawyer, and the wrongdoing acknowledged. It must be acknowledged, too, that punishment is necessary, but not the extremity of capital punishment. This line of argument is hard to accomplish when it is accompanied by a nuanced plea for mitigation on account of deprivation, abuse and hardship, for it is too painfully obvious, that the victim lies dead, and the convicted is alive and sentient.

Once on Death Row, all autonomy is removed from the condemned men and women, and they enter a harsh, isolating and traumatizing environment, where the State works tirelessly toward achieving their execution. Death Row inmates spend up to 30 years in cells that are 6 feet by 9 feet. More than 50 per cent of them are from ethnic minority groups. Conditions are inhumane – no natural light, poor ventilation, little association with other prisoners or visitors, restricted space, and leg irons as routine. Handcuffs, waist-chains and shackles are used if a prisoner leaves the cell. The brutality is unimaginable: this is a studied and practiced way of treating people as trash. The United Nations Standard Minimum Rules for the Treatment of Prisoners, to which the U.S. is a signatory, is violated at every turn, and there is no redress. Insanity amongst prisoners is frequent, as a result of the harsh physical treatment, and because of the psychological torture of waking up every day, year after year, knowing that the State is striving with every means in its power to take each life.

In the state of Florida, a prisoner on Death Row becomes "warrant eligible", once he has been through his appeals and the clemency proceedings. This means that the governor can sign his death warrant and set an execution date. The selection process is arbitrary – a prisoner might have been eligible for many years, but the governor may jump ahead to someone more recently arrived. Ronald W. Clark, Jr., describes the torment thus:

> Your only hope is that the governor's hit squad doesn't select you next. You sit there, like a trapped animal, in a 9 ft. × 7 ft. cage, watching man after man selected before you, marched off to their death. After each execution, you worry, wondering with great trepidation, if you will be the next victim of Florida's machinery of legalized murder. The electronic doors will pop dozens of times each day to allow the officers to come into the wing. Each time the staff hits that buzzer, your stomach knots up and your heart races, wondering if they're coming for you! Is it your time to die or are they about to kill someone you've befriended over the years? Torture! This is torture, mental torture at its best! They torture you and your loved ones beyond an 8th Amendment violation. This is taking cruel and unusual punishment to a whole new level, because during the last seven days of your life you're going to have a man posted in front of your death-watch cage watching every move you make, documenting everything you do, from sleeping to using the toilet! You're reminded every minute of every day that you're not just going to die. No, death is easy! You're going to be strapped down to a gurney and killed; murdered, under the mantel of justice, as your fellow citizens watch on in an observation room. Yes, you may lay on that gurney strapped down for as much as an hour or two, knowing with sick apprehension, that your death, your killing, is at hand.[14]

Andrew Lee Jones had seven execution dates and just before each one there was a stay of execution granted, once just 20 hours beforehand.[15] Michael

---

14 This is an extract from an article called "Waiting To Die", submitted by Ronald W. Clark Jr. to *The Wing of Friendship*, the Summer 1918 quarterly newsletter of Lifelines, a charity which provides pen friends to men and women on Death Row.

15 Arriens, *Welcome to Hell*, 174. Andrew Lee Jones had a penfriend who flew to meet him for the first time a few days before his execution. When she returned home, there was a last letter from him, in which he hoped she had not been too upset by his death.

Lambrix entered Death Row in Florida in 1984, a self-confessed drifter and no-hoper. After losing a critical appeal 12 years later, he wrote this, an assertion of personal autonomy in a place designed to crush every independent aspect of a human being:

> It is not enough to condemn us. In society's demented state of moral consciousness, we must first be stripped of our humanity before being deprived of our life. To recognise our humanity is to create a reflection of their own inherent imperfection, as well as face the truth that they are taking a human life. But to make us less than human pacifies society's guilt. They don't kill any particular individual, but rather something less than an individual. But for each of us, there is a Yellow Brick Road. An escape from the reality of our condemnation. A place of solace and security. The adversity we suffer remains and continues to plague us; continues to rob us of the humanity and individuality we so desperately cling to. But as long as we each keep sight of our own Yellow Brick Road, we will deprive our captors and executioners of the theft of our humanity and stand strong in our inner strength. Not only to survive – but to overcome.[16]

Conclusion

A constructivist social science has a liberating part to play in disclosing the workings of institutions and, by shining a light into the darkest places of these institutions, showing the awesome power of institutions over individuals and groups. All of the men and women on Death Row are individuals, and yet most are all part of outcast groups, – the indigent, the damaged, the racially disadvantaged, and those who cannot afford proper legal representation. Social structure is one of the focal points for social scientific analysis, and social science can, and should, ask questions about these patterns of discrimination, and vehemently question the practice of denying the disadvantaged access to justice and fairness. It is hampered in this task by what I have referred to as the 'domestification' process, which is such a powerful, yet elusive, ideological tool in parts of a nation where it is not common to look to other cultures and societies for answers to problems. Why should a nation look outward, when

16 Arriens, *Welcome to Hell*, 207. This is an extract from a letter written by Michael Lambrix. In the same book, p. 233–237, he writes an extraordinary account of his life, and his spiritual journey, which I have no space to reproduce here.

looking inward is so very satisfying, and provides neat answers to the complexities of crime, punishment, and human rights? The death penalty survives because the very words used to describe it, and to advocate for it, are pleasing to so very many people. The long, drawn-out vengeance that it offers is satisfying to so many.

But the death penalty is a massive politico-ethical failure. It is unequivocally an arbitrary system of torture, a scar on the body politic. In some states in the U.S., it is an instrument that politicians use for their own ends, choosing to view Death Row prisoners as expendable. This chapter has tried to make their humanity more visible, and recognise their dignified autonomy at the last, as exemplified in this letter from Richard Townes to a friend, just before his execution:

> I love you for the goodness and genuineness I can feel within you as a person, which I found has had a very positive effect on me. Perhaps it is your tolerance and extreme understanding and compassion that allowed myself to grow the little inside that I needed to be able to feel real as a person – not just an image. Thank you for giving so much yet receiving so little in return.[17]

Bibliography

Arriens Jan, ed., *Welcome to Hell: Letters and Writings from Death Row*, Boston, MA: Northeastern University Press, 2015.

Elias, Norbert, *The Civilising Process, Vol. II, State Formation and Civilization*, Oxford: Blackwell 1982.

Foucault, Michel, *Discipline and Punish,* Harmondsworth: Penguin, 1973.

Garland, David, *Peculiar Institution: America's Death Penalty in an Age of Abolition,* (Cambridge, MA: The Belknap Press of Harvard University, 2010).

Larson, Doran, 'Why Scandinavian Prisons are Superior', *The Atlantic*, (2013): September.

Lifelines Quarterly Newsletter, Summer 1918, Bristol: Lifelines.

Medlicott, Diana, 'Condemned to Artifice and Prevented from Being a Pirate: How Prisoners Convicted of Terrible Crimes Recognise Themselves in Discourse', in R. Hamilton (editor), *This Thing of Darkness: Perspectives on Evil and Human Wickedness,* New York: Rodopi 2001.

17 Arriens, *Welcome to Hell*, 171. Richard Townes wrote this letter to his penfriend three days before his execution in Virginia on January 23rd, 1996.

Medlicott, Diana, *Surviving the Prison Place: Narratives of Suicidal Prisoners*, Aldershot: Ashgate, 2001.

Mennell, Stephen, An Exceptional Civilising Process, *Journal of Classical Sociology (2013)* Vol 9 (1).

Sarat, Austin, *When the State Kills: Capital Punishment and the American Condition*, Princeton: Princeton University Press, 2001.

U.S. Code, 18 USC 2340, accessed 8th March 2016 http://www.law.cornell.edu/uscode/uscprint.html.

CHAPTER 2

Torture's In/Visibility

Hedi Viterbo

Abstract

Using recent cases and images concerning three countries' involvement in torture – the United Kingdom, the United States, and Israel – this chapter illustrates the pitfalls of dominant conceptions of torture and seeks to open up alternative approaches. Currently, real-time photographic and video documentation of torture tend to be privileged over other images. This has facilitated the denial of certain evidence or aspects of torture, while drawing attention away from the broader social structures and policies that have given rise to the documented incidents. Among the factors left outside the frame are political and legal practices and mechanisms that keep torture hidden from detainees and the general public while rendering detainees hyper-visible to the state. This chapter examines some of these excluded factors and argues that visual materials such as sketches and re-enactment pictures possess the unique evidentiary potential for highlighting and problematising them. The proposed approach to torture images also requires a different understanding of the phenomenon of torture itself. Legal conceptions, in particular, disregard two crucial dimensions: these in/visibility-inducing forces and law's own violence. Following discussion of the former dimension, the chapter turns to the latter with a focus on law's complicity in the concealment, non-disclosure, and destruction of evidence of torture.

Keywords

invisibility – law – violence – visual evidence – the United States – the United Kingdom – Israel/Palestine

Representing Torture under the Prevailing Evidentiary Paradigm

In modern times, state torture has shifted from being a public spectacle to a hidden practice, as Michel Foucault famously described in *Discipline and Punish*. This has resulted, among other things, in a paucity of publicly available

 | DOI:10.1163/9789004408012_004

photographic or video evidence of state torture. In their absence, alternative representations have been produced, such as the following (see Figure 2.1):

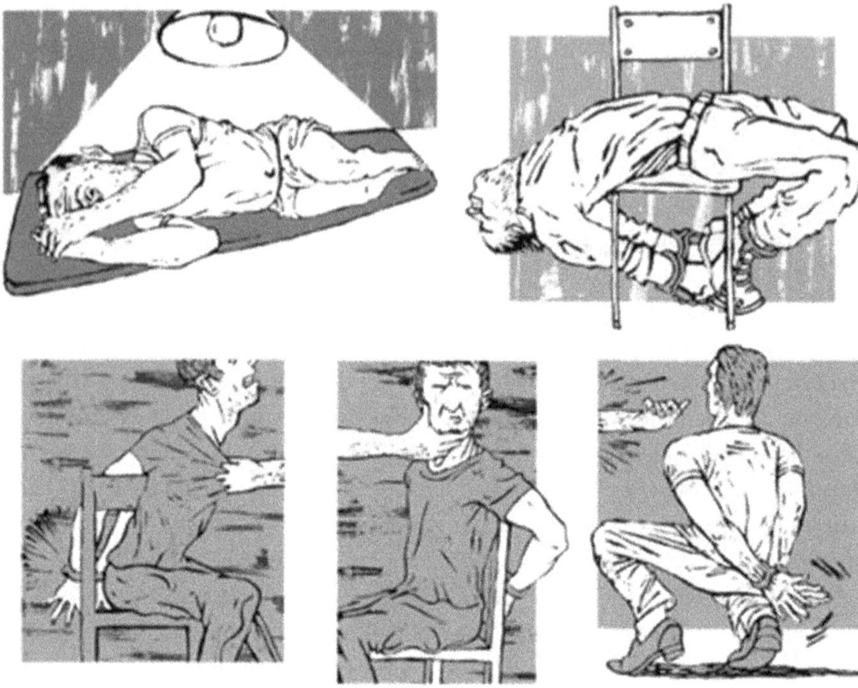

FIGURE 2.1 Coercive interrogation methods used by Israel

Featured prominently in *Absolute Prohibition* – a 2007 report by Israeli NGOs B'Tselem and Hamoked on Israel's torture and ill-treatment of Palestinian detainees – these images are meant to depict five coercive interrogation methods. Clockwise from top left: 1) sleep deprivation; 2) forcing interrogees into the so-called 'banana position', in which their body is arched backwards on a chair, sometimes while their handcuffs are shackled to their legs; 3) forcing interrogees to crouch on tiptoes in the so-called 'frog position' while their hands are cuffed behind their back, and pushing or beating them until they lose their balance; 4) sharply twisting their head; and, 5) suddenly pulling their body forward while they are handcuffed to the chair.

As John Tagg has observed in *The Burden of Representation*, in order to qualify as good evidence of torture, visual images are normally required to appear to be accurate, to create a sense of immediacy, of simply 'witnessing reality', while suppressing all traces of the mediation and representation at work. According to this dominant paradigm, photographs and videos are best suited for capturing the reality of torture, whereas sketches lack substantial evidentiary value. A possible exception is first-hand sketches drawn by torture victims or witnesses themselves,[1] as opposed to the above images, which, having been created by a professional illustrator (Ishai Mishory), may possess unique visual qualities but are inevitably condemned to a lesser evidentiary status. These drawings are therefore likely to be seen, at best, as illustrative of the 'real', non-visual evidence available to these NGOs: the verbal testimonies of Palestinian ex-detainees on which the report is based.

This prevailing evidentiary visual paradigm can unwittingly contribute to the downplaying, disregard, or even denial of certain evidence or aspects of state torture. The privileging of photographs and videos might facilitate, in their absence, the dismissal of other sources of information, which are looked upon as inferior evidence. Examples abound of this disregard of non-photographic evidence of state violence.[2] Perhaps the most graphic one is the situation of detainees tortured in US custody at Abu Ghraib prison in Iraq, the photographic images of which still stand out as probably the best known to have become public despite the invisibility shrouding torture. In comparison to the widespread global attention received following the publication of these images in 2004, earlier verbal accounts by NGOs and Iraqi ex-detainees about torture at Abu Ghraib were met with near silence in the public arena.

Moreover, this evidentiary paradigm tends to fixate on the specific details and incidents that torture photographs and videos are said to capture, while directing attention away from their political and institutional context, including the broader systematic use of torture. Indeed, many, especially in the United States, saw the Abu Ghraib images in isolation from the policies that brought them about, almost as if there was nothing to see outside the frame of the images. Accordingly, as Alfred McCoy has described in *Torture and Impunity*, there were only reprimands, disciplinary action, and courts-martial for

1 Examples of sketches drawn by detainees or their incarcerators can be found in: Save the Children (Sweden), *One Day in Prison – Feels Like a Year: Palestinian Children Tell Their Own Stories* (Stockholm, 2003), viewed on 29 May 2016, http://resourcecentre.savethechildren.se/sites/default/files/documents/2443.pdf; and, 'No Torture. No Exceptions', *Washington Monthly*, 2008, viewed on 29 May 2016, http://www.washingtonmonthly.com/features/2008/0801.torture.pdf.

2 For example: Noam Sheizaf, 'Beitunia Killings and the Media's Incredibly High Bar for Palestinian Stories', *+972 Magazine*, 21 May 2014, viewed on 29 May 2016, http://972mag.com/beitunia-killings-and-the-medias-incredibly-high-bar-for-palestinian-stories/91166/.

a few soldiers whom the Bush administration described as 'bad apples', while high-ranking officials have remained unaccountable.

This response is not unique to Abu Ghraib. In 2009, a video became public showing the torture of detainees in UK military custody in Iraq, one of whom, Baha Mousa, had died later that day. Only low-level British soldiers were prosecuted, and only the soldier seen abusing the detainees in the video was eventually convicted. He later insisted, however, that many other soldiers had also been violent toward detainees.[3]

Blindness to Torture's In/Visibility

A related pitfall of the dominant evidentiary paradigm is its blindness to the processes governing the visibility of torture, processes affecting both torture itself and the images that are said to capture it. In its modern form, state torture operates through a combination of invisibility and hyper-visibility. On the one hand, state torture is kept away from the public eye as well as from the eyes of its victims. Interrogational torture, specifically, owes much of its pervasiveness to taking place beyond public sight. This is why photographs and videos of torture, on the rare occasions that they do become publicly available, usually show events that occurred outside the interrogation room, as demonstrated by the abovementioned images of torture by US and UK forces in Iraq. Further, states seeking to keep torture invisible and deniable have increasingly deployed torture techniques that leave as few lasting physical marks as possible,[4] such as those depicted in the above sketches from the report on Palestinians in Israeli custody. Moreover, through blindfolding and hooding, state torture denies not only the general public but also detainees themselves access to potentially incriminating sights and information about the place of torture or the identity of the torturers.[5]

---

3 For further discussion of these and related aspects of the Abu Ghraib and Baha Mousa cases, see, respectively: Nicholas Mirzoeff, 'Invisible Empire: Visual Culture, Embodied Spectacle, and Abu Ghraib', *Radical History Review* 95 (2006): 21–44; and, Noel Whitty, 'Soldier Photography of Detainee Abuse in Iraq: Digital Technology, Human Rights and the Death of Baha Mousa', *Human Rights Law Review* 10.4 (2010): 689–714.

4 Darius Rejali, *Torture and Democracy* (Princeton and Oxford: Princeton University Press, 2007).

5 On the use of blindfolding and hooding by the United States and Israel, see, respectively, Human Rights Watch, *The Road to Abu Ghraib*, June 2004, https://www.hrw.org/sites/default/files/reports/usa0604.pdf; B'Tselem and Hamoked, *Kept in the Dark: Treatment of Palestinian Detainees in the Petah Tikva Interrogation Facility of the Israeli Security Agency*, Jerusalem, October 2010, viewed on 29 May 2016, https://www.btselem.org/download/201010_kept_in_the_dark_eng.pdf.

Central to concealing and denying the use of torture is secrecy. UK authorities have thus repeatedly refused, on various grounds, to release requested information on the government's involvement in torture overseas. British officials also ensured that any information on the matter was redacted from the 2014 US Senate Select Committee on Intelligence report on the use of torture by the Central Intelligence Agency (CIA).[6] As for the CIA itself, it has operated so-called 'black sites' – secret detention and interrogation facilities overseas.[7] Israel likewise concealed the existence of one of its interrogation and detention facilities, Facility 1391, until 2002.[8] Also in pursuit of secrecy, in 2009, the then US President, Barack Obama, tried to prevent the publication of previously unpublished photographs and videos of torture at Abu Ghraib. The following year, journalists were also expelled from Guantánamo for publishing the name of a witness testifying on interrogation techniques.[9]

On the other hand, in many cases detainees are photographed or videotaped during their torture, thus becoming hyper-visible to the state and its extensions. The CIA, for example, has videotaped many of its interrogations of terrorism suspects, and also reportedly photographed detainees who were stripped nude while awaiting interrogation – a form of sexual humiliation often used in tandem with other torture methods. To ensure this hyper-visibility operates only on the state's terms, however, the CIA eventually destroyed interrogation videotapes it considered a potential security risk, and keeps classified the photographs of naked detainees.[10] Similarly, on several known occasions, Israeli armed forces videotaped their abuse of Palestinians and later destroyed

6 Ruth Blakeley and Sam Raphael, 'British Torture in the "War on Terror"', *European Journal of International Relations* 23.2 (2017): 243–266.

7 Leila Sadat, 'Ghost Prisoners and Black Sites: Extraordinary Rendition under International Law', *Case Western Reserve Journal of International Law* 37.2 (2006): 309–342.

8 Laleh Khalili, *Time in the Shadows: Confinement in Counterinsurgencies* (Stanford: Stanford University Press, 2013).

9 See, respectively: Constitution Project, *Report of the Task Force on Detainee Treatment*, Washington DC, 16 April 2013, viewed on 29 May 2016, https://www.opensocietyfoundations.org/sites/default/files/constitution-project-report-on-detainee-treatment_0.pdf; Jeremy W. Peters, 'Pentagon Reinstates Banned Guantánamo Reporter', *New York Times*, 9 July 2010, viewed on 29 May 2016, https://mediadecoder.blogs.nytimes.com/2010/07/09/pentagon-reinstates-banned-guantanamo-reporter/.

10 Mark Denbeaux, et al., 'Captured on Tape: Interrogating and Videotaping of Detainees at Guantánamo', *Seton Hall Law Review* 41 (2011): 1307–1317; Spencer Ackerman, 'CIA Photographed Detainees Naked before Sending Them to Be Tortured', *The Guardian*, 28 March 2016, viewed on 29 May 2016, http://www.theguardian.com/us-news/2016/mar/28/cia-photographed-naked-detainees.

the videos. One such incident, in 2002, involved the fatal abuse of one Palestinian and the injury of three others in the West Bank. In that case, the abusive soldiers destroyed not only the videos but also other potentially incriminating evidence, in addition to coordinating their stories and threatening a comrade who had decided to testify against them."[11]

Rethinking Visual Evidence of Torture

Within the dominant evidentiary paradigm, then, torture images induce blindness. In so doing, they unwittingly contribute to dismissals of torture allegations. It is therefore imperative not only to seek more photographic and video documentation of torture, but also to bring into question torture's in/visibility. Notwithstanding its value, greater visibility – in the form of more photographic and video evidence of torture – might in some respects only further conceal state torture. This is borne out by some commentators, such as Stephen Eisenman in *The Abu Ghraib Effect*, who have questioned how much of an impact such photographs and videos actually have.

The supposed evidentiary deficiency of non-photographic torture images, such as the above sketches, both reflects and is a product of the specific evidentiary visual paradigm currently dominating the social imagination. Alternative ways of looking and thinking may open up other possibilities for engaging with the socio-political complexities and implications of state torture. This is not a matter of the intentions behind torture images, as images have a life of their own apart from any intentions those who produced them may or may not have had. Nor is it a matter of if and how some viewers can or will react to visual representations of torture, because if certain responses are considered improbable or counter-intuitive, then visual intuition can be reinvented by challenging the dominant cultural and social assumptions surrounding such images.

Giorgio Agamben has argued in *Remnants of Auschwitz* that testimony derives its evidentiary value precisely from what it is missing. Others, such as Nicholas Mirzoeff in *The Right to Look*, have emphasised the need to look for such seeming absences and make something out of them. Perhaps, then, it is precisely the non-realistic quality of sketches such as those above that carries

11 Hedi Viterbo, 'Seeing Torture Anew: A Transnational Reconceptualization of State Torture and Visual Evidence', *Stanford Journal of International Law* 50.2 (2014): 281–317.

its own evidentiary potential. Their non-realism constitutes these images as evidence of the lack of other visual representations (photographs or videos) of the depicted torture methods, and thereby potentially calls to mind the mechanisms and practices keeping torture out of public sight. Such sketches can thus function as meta-pictures, a term coined by W.J.T. Mitchell in *Picture Theory* to denote representations of the representation process itself.

Besides sketches, another type of metapicture is re-enactment photographs and videos. In such images, the re-enactors are sometimes the actual former detainees that were subjected to torture, as in the following image (see Figure 2.2):

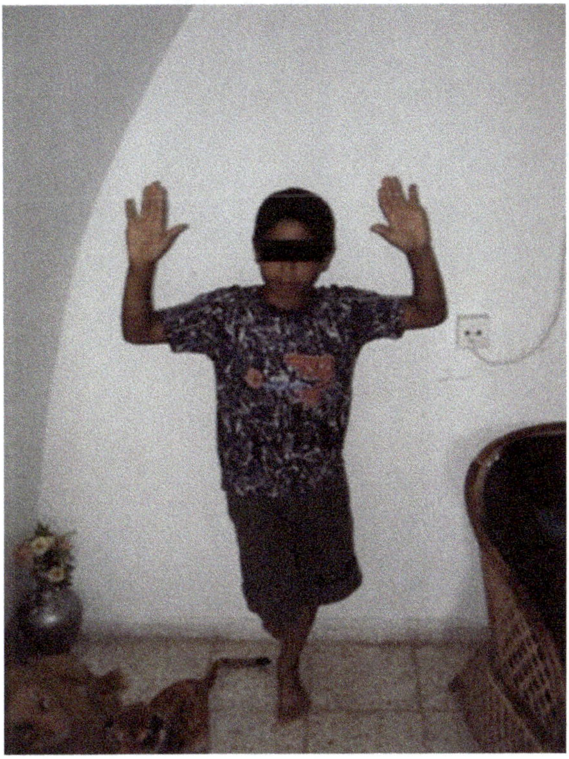

FIGURE 2.2 Ezzat: re-enacting stress position
© 2009 DEFENCE FOR CHILDREN INTERNATIONAL. PALESTINE. USED WITH PERMISSION.

Appearing in a report the Palestine Section of NGO Defence for Children International published in 2009, *Palestinian Child Prisoners*, the photograph shows a 10-year-old Palestinian re-enacting the stress position to which he alleges Israeli soldiers subjected him: standing on one foot and lifting his hands in the air for about half an hour.

In other cases, it is not the former detainees but others who participate in the re-enactment. Examples include the photograph, pervasive on the internet, showing protesters demonstrating the use of water-boarding on a volunteer in front of the US Justice Department building; or the video famously released in 2013 by NGO Reprieve showing American hip-hop artist Yasiin Bey (formerly known as Mos Def) undergoing force feeding.[12] A still from the latter appears below (see Figure 2.3):

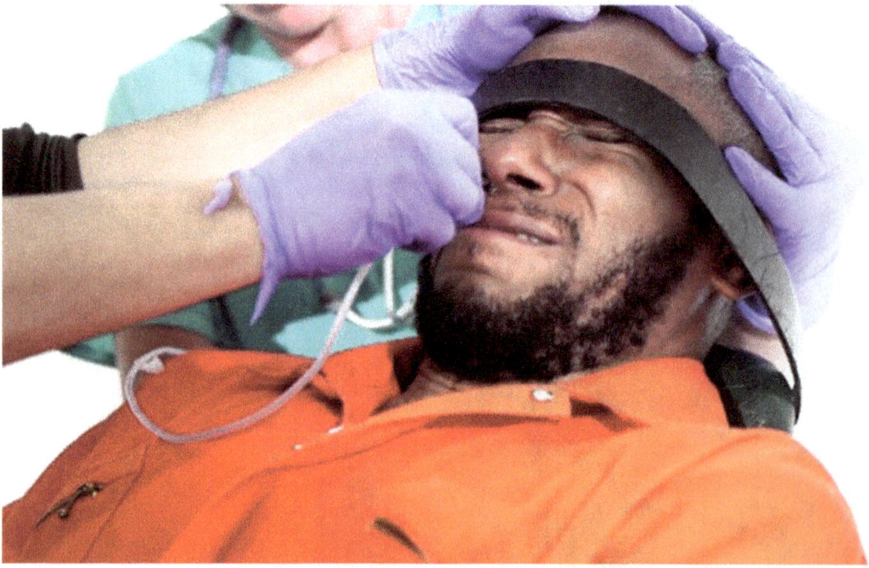

FIGURE 2.3 Yasiin Bey's force-feeding re-enactment

Treating such images as nothing but a simulation of real events misses their capacity to serve as a reminder of the very real reason for resorting to re-enactment in the first place: the social, political, and legal forces that

12 For the photograph of protesters re-enacting water-boarding and the video of Yasiin Bey / Mos Def see, respectively, David Boroff, 'Wisconsin Man Waterboarded Girlfriend to Find Out Whether She Was Cheating on Him', *New York Daily News*, 2 March 2016, viewed on 29 May 2016, http://www.nydailynews.com/news/crime/wis-man-waterboarded-gal-pal-find-cheated-article-1.2550123; Ben Ferguson, 'When Yasiin Bey Was Force-Fed Guantánamo Bay-Style – Eyewitness Account', *The Guardian*, 9 July 2013, viewed on 29 May 2016, http://www.theguardian.com/world/shortcuts/2013/jul/09/yasiin-bey-force-fed-guantanomo-bay-mos-def.

render state torture invisible, such as the exclusion of non-state media from torture sites.

In contrast to the sort of photographs and videos that are likely to be regarded as good visual evidence, then, the evidentiary power of alternative images of torture, such as sketches or re-enactment pictures, largely lies not in their providing what appears to be an unmediated record of state torture. Rather, it is through their mediating character that these images gain a particular evidentiary potential of not only documenting torture incidents, but also exposing how torture is mediated both by state efforts to control its visibility and by non-governmental processes of representation. Instead of providing viewers with a sense of simply witnessing state torture, these images thus potentially intimate the seemingly invisible and all-too-often unexamined representation at work.

However, this does not simply mean that supposedly weak visual evidence actually brings torture to light whereas privileged visual evidence fails to do so. First, the task at hand is not simply to have sketches, re-enactment photographs, or other images supposedly expose the invisibility surrounding torture, but rather to make the relationship between visibility and invisibility the object of inquiry. To this end, an investigation is needed into how, and to what effect, visual representations of torture, like the practice of torture itself, oscillate between invisibility and visibility, between absence and presence. Second, the aim of the critical aesthetic proposed here is not to resolve torture's in/visibility – as if such were at all possible – by proclaiming the invisible to be visible, or vice versa. Instead, by foregrounding and subjecting to inquiry oscillations between visibility and invisibility, the aim is to suggest how alternative ways of looking and thinking might potentially deconstruct the visible/invisible binary itself.

Law's Complicity in Concealing Torture

Engaging differently with visual representations of torture, as suggested so far, requires and depends on a different understanding of the phenomenon of torture itself. This includes calling into question the way in which international law defines torture specifically, and state violence generally. In the UN Convention Against Torture, torture is defined as 'any act by which severe pain or suffering, whether physical or mental, is intentionally inflicted on a person' by a state official for purposes specified in the Convention. This, the Convention notes, 'does not include pain or suffering ... inherent in or incidental to lawful sanctions'. The four Geneva Conventions, the chief international legal treaties governing humanitarian treatment in war, all prohibit 'violence to life and persons' more broadly, listing 'in particular … cruel treatment and torture'.

These legal definitions render invisible two crucial aspects of torture and violence. The first, discussed thus far, is the in/visibility through which state violence operates, is experienced, and is made possible. By focusing on physical and psychological harm, these legal texts leave out of the picture this violent in/visibility. The second overlooked and excluded aspect, to which this chapter now turns, is law's own violence, which these legal texts deny by counterposing law to violence.

Indeed, a socially prevalent notion, particularly in countries that perceive or present themselves as law-abiding democracies, is that law stands in opposition to violence. According to this view, acts such as murder or property offences are illegal and violent, whereas acts carried out under legal auspices, such as the death penalty or unjust taxation, are effectively perceived as non-violent.[13] What thus remains out of sight is the violence that is part and parcel of law's routine operation, the violence in each and every legal decision and interpretation, the violence through which law realises itself and on which it therefore depends, the violence that law occasions or justifies.[14]

Part of law's function is to deny its own violence, as evidenced by the above legal texts. As Elizabeth Grosz has put it, following Jacques Derrida, what 'we sometimes name the law, right, or reason' is 'a kind of counter-violence whose violence consists in the denial of violence. … This is a violence that describes and … structures itself as lawful, and thus beyond or above violence, that which judges violence'.[15] To justify existing legal definitions of violence and torture by appealing to 'ordinary language' or 'common sense' is simply to obfuscate this inherent violence,[16] as well as the violence of language at large.[17]

This violence of law includes or manifests itself in law's culpability for state torture. The contribution of lawyers, legal arguments, and legal institutions to

---

13 Robert Paul Wolff, 'On Violence', *Journal of Philosophy* 66.19 (1969): 601–616.

14 Walter Benjamin, 'Critique of Violence'. *Reflections: Essays, Aphorisms and Autobiographical Writings*, edited by Peter Demetz (Berlin: Schoken, 1986), 277–300; Robert M. Cover, 'Violence and the Word', *Yale Law Journal* 95 (1986): 1601–1629; Jacques Derrida, 'Force of Law: The Mystical Foundations of Authority', *Cardozo Law Review* 11.5–6 (1990): 919–1046.

15 Elizabeth Grosz, 'The Time of Violence: Deconstruction and Value', *Cultural Values* 2.2 (1998): 190–205.

16 For a similar criticism, see Yves Winter, 'Violence and Visibility', *New Political Science* 34.2 (2012): 195–202.

17 On the violence of language, see, for example, Slavoj Žižek, *Violence: Six Sideways Reflections* (New York: Picador, 2008).

shaping and legitimising torture has been extensively studied. Books such as Alan Clarke's *Rendition to Torture*, Karen Greenberg's *Rogue Justice*, and John Parry's *Understanding Torture*, as well as a growing number of articles, have all tackled this issue. A few representative examples would therefore suffice. In 2005, faced with accusations that his government was an accomplice to extrajudicial abductions of terror suspects by the United States, the then British Prime Minister, Tony Blair, insisted on portraying such abductions as legal: 'all I know', he said, 'is that we should keep within the law ... and the notion that I, or the Americans, ... condone torture ... is completely ... out of order ... I have absolutely no evidence to suggest that anything illegal has been happening here at all'.[18] Somewhat similarly, in the United States, the infamous 'torture memos' drafted by lawyers in the George W. Bush administration interpreted the law so as to permit the use of highly controversial interrogation methods.[19] A 1999 ruling of the Israeli Supreme Court likewise granted impunity, under the 'necessity defence', to interrogators who use torture in so-called 'exceptional circumstances'.[20] As these examples illustrate, it is through legal rhetoric and mechanisms that all three countries have sought to legitimise their contentious actions.

Not only violence but in/visibility too is integral to law. As Costas Douzinas puts it, to some extent the law is 'always somewhere else, in the next room, deferred and unseen ... a sign of the transcendent apprehended in its absence', just as in Franz Kafka's parable "Before the Law"'.[21] Embodying this interplay between law, violence and in/visibility is Lady Justice – the allegorical personification of law, typically depicted as blindfolded and holding a sword. Though her blindfold has come to symbolise the supposed impartiality of law, in the past it actually represented law's inability to deliver justice.[22] Lady Justice's

18 'Rendition: Blair in Quotes', *BBC*, 19 January 2006, viewed on 29 May 2016, http://news.bbc.co.uk/2/hi/uk_news/politics/4627360.stm.

19 David Cole ed., *The Torture Memos: Rationalizing the Unthinkable* (New York and London: New Press, 2009).

20 Ardi Imseis, '"Moderate" Torture on Trial: Critical Reflections on the Israeli Supreme Court Judgment Concerning the Legality of the General Security Service Interrogation Methods', *Berkeley Journal of International Law* 19.2 (2001): 328–349.

21 Costas Douzinas, 'Prosopon and Antiprosopon: Prolegomena for a Legal Iconology' in *Law and the Image: The Authority of Art and the Aesthetics of Law*, ed. Costas Douzinas and Lynda Nead (Chicago and London: University of Chicago Press, 1999), 36–68.

22 Erwin Panofsky, *Studies in Iconology: Humanistic Themes in the Art of the Renaissance* (New York: Harper & Row, 1972); M. Jay, 'Must Justice Be Blind?: The Challenge of Images to the Law' in *Law and the Image: The Authority of Art and the Aesthetics of Law*, ed. Costas Douzinas and Lynda Nead (Chicago and London: University of Chicago Press, 1999), 19–35.

blindfold could thus be interpreted as illustrating a bidirectional flow of invisibility and concealment: the blindfold keeps those who are the targets of law's violence out of its sight or interest, while at the same time preventing them from knowing the direction of law's gaze.

This close relationship between law, violence and in/visibility, however, tends to go unnoticed. In the present context, as mentioned above, law's complicity in shaping and legitimising the use of torture has been the subject of considerable scholarship. Far less attention has been paid to law's involvement, by commission or omission, in the concealment, non-disclosure or destruction of potentially incriminating evidence of torture, and also in granting impunity to alleged torturers. Examples abound of this overlooked function of law regarding each of the three countries under examination.

A recent case in point was the enactment of the Justice and Security Act in the United Kingdom in 2013, by which so-called 'closed material procedures' (previously reserved for special tribunals and specific circumstances) were extended into the main civil courts. Such procedures might ultimately allow the government to withhold information from the public about its involvement in torture, by enabling it to introduce evidence that is only seen by the judge and security-cleared 'special advocates'.[23] Indeed, this procedure was put to use in 2017, in a Pakistani national's claim concerning his extrajudicial abduction by British forces from Iraq to a US detention facility in Afghanistan, where he had been allegedly tortured and incarcerated without charge for over ten years.[24] Secret evidence that is not disclosed to the claimants was also the basis on which, in 2016, the High Court struck out claims by a group of Libyan men who had been kidnapped by the UK and US security agencies on behalf of Libya. That same year, the Crown Prosecution Service announced it would not bring charges against British officials, despite confirming their involvement in these kidnappings. Claims by more than 600 Iraqis that British soldiers had physically mistreated them were also dismissed in 2016 by the Supreme Court, which held that they had been brought too late.[25]

---

23 Owen Bowcott, 'What are Secret Courts and What Do They Mean for UK Justice?', *The Guardian*, 14 June 2013, viewed on 29 May 2016, https://www.theguardian.com/law/2013/jun/14/what-are-secret-courts.

24 Owen Bowcott, 'Rendition: Government Evidence to Be Heard in Secret in UK for First Time', *The Guardian*, 23 March 2017, viewed on 31 March 2017, ,https://www.theguardian.com/world/2017/mar/23/rendition-government-evidence-heard-secret-uk-first-time.

25 See, respectively: Ian Cobain, 'Libyan Men's Claims against UK over Gaddafi Cooperation Thrown Out', *The Guardian*, 15 April 2016, viewed on 29 May 2016, http://www.theguardian.com/law/2016/apr/15/libyan-claims-against-uk-thrown-out-secret-evidence; 'UK Was Involved in Libyan Torture Flights and Politicians Knew, Say British Prosecutors',

In the United States, in 2015, the Supreme Court left in place a lower court decision, which, for 'national security' reasons, exempted videotapes of the allegedly coercive interrogation and confinement of a Guantánamo detainee from disclosure under the Freedom of Information Act. A year later, in another case, defence attorneys at Guantánamo's military commissions accused the trial judge and prosecution team of involvement in the destruction of favourable defence evidence; due to rules of war-court secrecy, however, the defence lawyers were prevented from publicly describing the evidence that had allegedly been destroyed. Further, high-ranking CIA officials who, as described above, pushed for the destruction of interrogation videotapes, also did so in reliance on law, interpreting it as placing no obligation to retain such materials. In addition, under President Obama, and in line with his public statements, the Justice Department ruled out prosecutions over torture in US custody as long as interrogators followed legal advice. It also closed without charges the only two cases that were under investigation, one of which had resulted in the death of an Iraqi detainee at Abu Ghraib.[26] Furthermore, no less than its ally across the Atlantic, the United States has frequently used secret evidence in proceedings described as involving national security matters. Review of continued detention at Guantánamo, for example, can be based on secret evidence undisclosed to the detainee.[27]

Legally sanctioned secrecy and impunity have been equally central to preventing public access to potential evidence of state torture in Israel/Palestine. Judicial review of so-called 'administrative detention' – incarceration without

---

Reprieve, 9 June 2016, viewed on 10 June 2016, http://www.reprieve.org.uk/press/uk-was-involved-in-libyan-torture-flights-and-politicians-knew-say-british-prosecutors/; Owen Bowcott, 'Court Rejects 600 Iraqis' Claims of Mistreatment by UK Soldiers', *The Guardian*, 12 May 2016, viewed on 10 June, 2016, http://www.theguardian.com/law/2016/may/12/court-rejects-600-iraqis-claims-of-mistreatment-by-uk-soldiers.

26 Lawrence Hurley, 'Supreme Court Rejects Cases on Guantanamo Detainee Treatment', *Fiscal Times*, 9 March 2015, viewed on 29 May 2016, http://www.thefiscaltimes.com/latestnews/2015/03/09/US-top-court-blocks-suit-Syrian-former-Guantanamo-detainee; Carol Rosenberg, 'Alleged 9/11 Plotter's Lawyers Ask Prosecutor, Judge to Quit Trial over Destruction of Evidence', *Miami Herald*, 11 May 2016, viewed on 29 May 2016, http:// www.miamiherald.com/news/nation-world/world/americas/guantanamo/article77015207.html; Douglas Cox, 'Burn After Viewing: The CIA's Destruction of the Abu Zubaydah Tapes and the Law of Federal Records', *National Security Law* 5 (2011): 131– 177; Glenn Greenwald, 'Obama's Justice Department Grants Final Immunity to Bush's CIA Torturers', *The Guardian*, 31 August 2012, viewed on 29 May 2016, http://www.guardian.co.uk/commentisfree/2012/aug/31/obama-justice-department-immunity-bush-cia-torturer.

27 Kent Roach, *The 9/11 Effect: Comparative Counter-Terrorism* (New York: Cambridge University Press, 2011).

charge or trial for indeterminate periods of time – is unbound by the regular rules of evidence, is based on secret evidence, and is held behind closed doors.[28] More than 300 appeals were submitted to the Supreme Court against extensions of such detention between the years 2000 and 2010, but not a single one resulted in a release order or in a rejection of the secret evidence.[29] Relatedly, of over 800 complaints of torture or abuse by Israeli interrogators that were submitted to the State Attorney's Office between 2001 and 2014, none resulted in a criminal investigation.[30] In addition, in 2013, Israel's Supreme Court dismissed a petition requiring the videotaping of interrogations of suspected 'security offenders', the overwhelming majority of whom are Palestinians.[31]

The violence-law-in/visibility triad is by no means unique to torture, though it plays out differently in different types and contexts of state violence. A case in point is extrajudicial state-sponsored assassinations, commonly referred to as 'targeted killings'. Despite being conducted much more visibly than state torture, this form of state violence, too, relies on various and partly intertwined invisibilities and legalities. For instance, in late 2017, a UK tribunal dismissed a request for information about the legal advice given to the Prime Minister prior to drone strikes in Syria.[32] During the Obama presidency, and somewhat similarly to the 'torture memos' of the George W. Bush administration, the US Justice Department's Office of Legal Counsel authored a memo approving the extrajudicial drone killing of an American citizen in Yemen, which subsequently became the subject of several lawsuits.[33] In Israel, in 2008, a former soldier

---

28 Amnesty International, *Starved of Justice: Palestinians Detained Without Trial by Israel*, London, 2012, viewed on 29 May 2016, https://www.amnestyusa.org/wp-content/uploads/2017/04/israelopt_starved_of_justice_060612-final.pdf.

29 Shiri Krebs, 'Lifting the Veil of Secrecy: Judicial Review of Administrative Detentions in the Israeli Supreme Court', *Vanderbilt Journal of Transnational Law* 45.3 (2012): 639–703.

30 Public Committee Against Torture in Israel, *Prosecutorial Indifference: Systematic Failures in the Investigation of Soldier Violence Against Detainees in the Occupied Palestinian Territories*, Jerusalem, June 2014, viewed on 29 May 2016, http://stoptorture.org.il/wp-content/uploads/2015/10/%D7%91%D7%9E%D7%A6%D7%97-%D7%9C%D7%90-%D7%A0%D7%97%D7%95%D7%A9%D7%94-%D7%92%D7%A8%D7%A1%D7%94-%D7%91%D7%90%D7%A0%D7%92%D7%9C%D7%99%D7%AAProsecutorial-Indifference.pdf.

31 Viterbo.

32 Owen Bowcott, 'Tribunal Rules Against Total Secrecy over UK Drone Strikes', *The Guardian*, 4 January 2018, viewed 4 January 2018, https://www.theguardian.com/politics/2018/jan/04/tribunal-rules-against-total-secrecy-over-uk-drone-strikes.

33 Department of Justice – Office of Legal Counsel, *Memorandum for the Attorney General Re: Applicability of Federal Criminal Laws and the Constitution to Contemplated Lethal Operations Against Shaykh Anwar al-Aulaqi* (Washington DC, July 16 2010, viewed on 29 May 2016), https://www.nytimes.com/interactive/2014/06/23/us/23awlaki-memo.html.

leaked to a journalist classified military documents suggesting that the military had violated a Supreme Court ruling by assassinating Palestinians who could have been arrested. But the Israeli Attorney General refused to investigate the reportedly unlawful killing. The only ones prosecuted and convicted were the former soldier and journalist who revealed this information.[34] In a rhetorical twist, the judgments on their matter depicted their unauthorised possession of military documents as a 'ticking bomb' – a factually and ethically questionable scenario that is often invoked to justify state torture. These examples illustrate the importance of tackling the centrality of in/visibility and law not only to torture but to violence in general.

Reconceptualising Violence

What the discussion so far suggests is that the language and practice of law on the one hand, and the prevalent evidentiary paradigm regarding images of violence on the other, parallel and complement one another. Both of them mask and potentially facilitate central dimensions of state violence, and in this regard they themselves are inextricable from the violence they conceal. Therefore, just as alternative ways of engaging with visual representations of torture are needed, so should state violence – and torture in particular – be delineated beyond law's restrictive perimeters; that is, beyond the equation of violence with illegal and discrete incidents of physical or psychological harm.

Alternative conceptions may lay bare some of the dimensions of torture and violence that legal definitions conceal or deny. In his widely cited writing on the subject, Johan Galtung maintains that 'violence is present when human beings are being influenced so that their actual somatic and mental realizations are below their potential realizations. ... Violence is here defined as the cause of the difference ... between what could have been and what is'. Accordingly, beyond illegal acts of interpersonal physical or psychological violence, there is also 'structural or indirect' violence 'built into the [social] structure', which 'shows up as unequal power and consequently as unequal life chances'. Interrelated with this structural violence are 'those aspects of culture, ... [such as] religion and ideology, language and art, ... science' – and,

34 Dan Izenberg, 'Stamp of Approval from Attorney-General', *Jerusalem, Post* 13 April 2010, viewed on 29 May 2016, http://www.jpost.com/Israel/Stamp-of-approval-from-attorney-general.

arguably, law – 'that can be used to justify or legitimize direct or structural violence'.[35]

Michel Foucault, Rob Nixon, and Pierre Bourdieu have each conceptualised violence along somewhat similar or related lines. For Foucault, a 'relationship of violence' is one that 'acts upon a body or upon things; it forces, it bends, it breaks on the wheel, it destroys, or it closes the door on all possibilities. … [I]f it comes up against any resistance, it has no other option but to try to minimize it'.[36] Meanwhile, Nixon describes 'slow violence' as 'violence that occurs gradually and out of sight, a violence of delayed destruction that is … typically not viewed as violence at all … a violence that is neither spectacular nor instantaneous, but rather incremental and accretive'.[37] Bourdieu, in turn, developed the concept of 'symbolic violence'. As the translator of one of his writings clarifies, this violence concerns 'the imposition of … symbolic representations (languages, conceptualizations, portrayals), on recipients who have little choice about whether to accept or reject them'. Occurring in and beyond law, such violence involves the use of 'symbolic capital', such as '[a]uthority [and] knowledge'.[38]

Contrary to the view represented by the Convention Against Torture and the Geneva Conventions, then, violence can be understood as often operating indirectly; non-personally; in close relation to, rather than in disregard of, the law; and in ways irreducible to, though still potentially leading to, physical or psychological harm. By going beyond legal definitions, as well as beyond the prevailing evidentiary-visual paradigm, alternative and critical engagements with the interplay of state violence, law, and in/visibility can be developed, in and beyond the torture context.

Bibliography

Ackerman, Spencer. 'CIA Photographed Detainees Naked before Sending Them to be Tortured'. *The Guardian*, 28 March 2016. Viewed on 29 May 2016, http://www.theguardian.com/us-news/2016/mar/28/cia-photographed-naked-detainees.

35 Johan Galtung, 'Peace, and Peace Research', *Journal of Peace Research* 6.3 (1969): 167–191; Johan Galtung, 'Cultural Violence', *Journal of Peace Research* 27.3 (1990): 291–305.

36 Michel Foucault, 'The Subject and Power', *Critical Inquiry* 8.4 (1982): 777–795.

37 Rob Nixon, *Slow Violence and the Environmentalism of the Poor* (Cambridge, MA and London: Harvard University Press, 2011).

38 Richard Terdiman, 'Translator's Introduction to Pierre Bourdieu's The Force of Law: Toward a Sociology of the Juridical Field', *Hastings Law Journal* 38 (1987): 805–813.

Agamben, Giorgio. *Remnants of Auschwitz: The Witness and the Archive.* New York: Zone Books, 1999.

Amnesty International. *Starved of Justice: Palestinians Detained Without Trial by Israel.* London, 2012. Viewed on 29 May 2016, https://www.amnestyusa.org/wp-content/uploads/2017/04/israelopt_starved_of_justice_060612-final.pdf.

Benjamin, Walter. 'Critique of Violence'. *Reflections: Essays, Aphorisms and Autobiographical Writings,* edited by Peter Demetz, Berlin: Schoken, 1986, 277–300.

Blakeley, Ruth and Sam Raphael. 'British Torture in the "War on Terror"'. *European Journal of International Relations* 23.2 (2017): 243–266.

Boroff, David. 'Wisconsin Man Waterboarded Girlfriend to Find Out Whether She Was Cheating on Him'. *New York Daily News,* 2 March 2016. Viewed on 29 May 2016, http://www.nydailynews.com/news/crime/wis-man-waterboarded-gal-pal-find-cheated-article-1.2550123.

Bourdieu, Pierre. *Language and Symbolic Power.* Oxford: Polity Press, 1991.

Bowcott, Owen. 'What are Secret Courts and What Do They Mean for UK Justice?' *The Guardian,* 14 June 2013. Viewed on 29 May 2016, https://www.theguardian.com/law/2013/jun/14/what-are-secret-courts.

Bowcott, Owen. 'Court Rejects 600 Iraqis' Claims of Mistreatment by UK Soldiers'. *The Guardian,* 12 May 2016. Viewed on 10 June 2016, http://www.theguardian.com/law/2016/may/12/court-rejects-600-iraqis-claims-of-mistreatment-by-uk-soldiers.

Bowcott, Owen. 'Rendition: Government Evidence to Be Heard in Secret in UK for First Time'. *The Guardian,* 23 March 2017. Viewed on 31 March 2017, https://www.theguardian.com/world/2017/mar/23/rendition-government-evidence-heard-secret-uk-first-time.

Bowcott, Owen. 'Tribunal Rules Against Total Secrecy over UK Drone Strikes'. *The Guardian,* 4 January 2018. Viewed 4 January 2018, https://www.theguardian.com/politics/2018/jan/04/tribunal-rules-against-total-secrecy-over-uk-drone-strikes.

B'Tselem and Hamoked. *Absolute Prohibition: The Torture and Ill-treatment of Palestinian Detainees.* Jerusalem, May 2007. Viewed on 29 May 2016. http://www.btselem.org/download/200705_utterly_forbidden_eng.pdf.

B'Tselem and Hamoked. *Kept in the Dark: Treatment of Palestinian Detainees in the Petah Tikva Interrogation Facility of the Israeli Security Agency.* Jerusalem, October 2010. Viewed on 29 May 2016. https://www.btselem.org/download/201010_kept_in_the_dark_eng.pdf.

Clarke, Alan W. *Rendition to Torture.* New Brunswick: Rutgers University Press, 2012.

Cobain, Ian. 'Libyan Men's Claims against UK over Gaddafi Cooperation Thrown Out'. *The Guardian,* 15 April 2016. Viewed on 29 May 2016, http://www.theguardian.com/law/2016/apr/15/libyan-claims-against-uk-thrown-out-secret-evidence.

Cole, David (editor). *The Torture Memos: Rationalizing the Unthinkable.* New York and London: New Press, 2009.

Constitution Project. *Report of the Task Force on Detainee Treatment*. Washington DC, 16 April 2013. Viewed on 29 May 2016. https://www.opensocietyfoundations.org/sites/default/files/constitution-project-report-on-detainee-treatment_0.pdf.

Cover, Robert M. 'Violence and the Word'. *Yale Law Journal* 95 (1986): 1601–1629.

Cox, Douglas. 'Burn After Viewing: The CIA's Destruction of the Abu Zubaydah Tapes and the Law of Federal Records'. *National Security Law* 5 (2011): 131–177.

Defence for Children International – Palestine. *Palestinian Child Prisoners: The Systematic and Institutionalised Ill-treatment and Torture of Palestinian Children by Israeli Authorities*. Jerusalem, June 2009. Viewed on 29 May 2016. https://d3n8a8pro7vhmx.cloudfront.net/dcipalestine/pages/1298/attachments/original/1433987832/DCIP_childprisoner_report.pdf?1433987832.

Denbeaux, Mark, Joshua Denbeaux, Matthew Darby, Adam Deutsch, Jennifer Ellick, David Gratz and Michael Ricciardelli, 'Captured on Tape: Interrogating and Videotaping of Detainees at Guantánamo', *Seton Hall Law Review* 41 (2011): 1307–1317.

Department of Justice – Office of Legal Counsel. *Memorandum for the Attorney General Re: Applicability of Federal Criminal Laws and the Constitution to Contemplated Lethal Operations Against Shaykh Anwar al-Aulaqi*. Washington DC, July 16 2010. Viewed on 29 May 2016. https://www.nytimes.com/interactive/2014/06/23/us/23awlaki-memo.html.

Derrida, Jacques. 'Force of Law: The Mystical Foundations of Authority'. *Cardozo Law Review* 11.5–6 (1990): 919–1046.

Douzinas, Costas. 'Prosopon and Antiprosopon: Prolegomena for a Legal Iconology'. *Law and the Image: The Authority of Art and the Aesthetics of Law*, edited by Costas Douzinas and Lynda Nead, Chicago and London: University of Chicago Press, 1999, 36–68.

Eisenman, Stephen F. *The Abu Ghraib Effect*. London: Reaktion Books, 2010.

Ferguson, Ben. 'When Yasiin Bey Was Force-Fed Guantnomo Bay-Style – Eyewitness Account', *Guardian*, 9 July, 2013. Viewed on 29 May 2016. http://www.theguardian.com/world/shortcuts/2013/jul/09/yasiin-bey-force-fed-guantanomo-bay-mos-def.

Foucault, Michel. *Discipline and Punish: The Birth of the Prison*. New York: Pantheon Books, translated by Alan Sheridan, 1978.

Foucault, Michel. 'The Subject and Power'. *Critical Inquiry* 8.4 (1982): 777–795.

Galtung, Johan. 'Cultural Violence'. *Journal of Peace Research* 27.3 (1990): 291–305.

Galtung, Johan. 'Peace, and Peace Research', *Journal of Peace Research* 6.3 (1969): 167–191.

Greenberg, Karen J. *Rogue Justice: The Making of the Security State*. New York: Crown, 2016.

Greenwald, Glenn. 'Obama's Justice Department Grants Final Immunity to Bush's CIA Torturers'. *The Guardian*, 31 August 2012. Viewed on 29 May 2016. http://www.guardian.co.uk/commentisfree/2012/aug/31/obama-justice-department-immunity-bush-cia-torturer.

Grosz, Elizabeth. 'The Time of Violence: Deconstruction and Value'. *Cultural Values* 2.2 (1998): 190–205.

Human Rights Watch. *The Road to Abu Ghraib*. June 2004. https://www.hrw.org/sites/default/files/reports/usa0604.pdf.

Hurley, Lawrence. 'Supreme Court Rejects Cases on Guantanamo Detainee Treatment'. *Fiscal Times*, 9 March 2015. Viewed on 29 May 2016. http://www.thefiscaltimes.com/latestnews/2015/03/09/US-top-court-blocks-suit-Syrian-former-Guantanamo-detainee.

Imseis, Ardi. '"Moderate" Torture on Trial: Critical Reflections on the Israeli Supreme Court Judgment Concerning the Legality of the General Security Service Interrogation Methods'. *Berkeley Journal of International Law* 19.2 (2001): 328–349.

Izenberg, Dan. 'Stamp of Approval from Attorney-General'. *Jerusalem Post* 13 April 2010. Viewed on 29 May 2016. http://www.jpost.com/Israel/Stamp-of-approval-from-attorney-general.

Jay, Martin. 'Must Justice Be Blind?: The Challenge of Images to the Law'. *Law and the Image: The Authority of Art and the Aesthetics of Law*, edited by Costas Douzinas and Lynda Nead. Chicago and London: University of Chicago Press, 1999.

Khalili, Laleh. *Time in the Shadows: Confinement in Counterinsurgencies*. Stanford: Stanford University Press, 2013.

Krebs, Shiri. 'Lifting the Veil of Secrecy: Judicial Review of Administrative Detentions in the Israeli Supreme Court'. *Vanderbilt Journal of Transnational Law* 45.3 (2012): 639–703.

McCoy, Alfred W. *Torture and Impunity: The U.S. Doctrine of Coercive Interrogation*. Madison: University of Wisconsin Press, 2012.

Mirzoeff, Nicholas. 'Invisible Empire: Visual Culture, Embodied Spectacle, and Abu Ghraib'. *Radical History Review* 95 (2006): 21–44.

Mirzoeff, Nicholas. *The Right to Look: A Counterhistory of Visuality*. Durham: Duke University Press, 2011.

Mitchell, W. J. T. *Picture Theory: Essays on Verbal and Visual Representation*. Chicago and London: Chicago University Press, 1995.

Nixon, Rob. *Slow Violence and the Environmentalism of the Poor*. Cambridge, MA and London: Harvard University Press, 2011.

'No Torture. No Exceptions'. *Washington Monthly*, 2008. Viewed on 29 May 2016. http://www.washingtonmonthly.com/features/2008/0801.torture.pdf.

Panofsky, Erwin. *Studies in Iconology: Humanistic Themes in the Art of the Renaissance*. New York: Harper & Row, 1972.

Parry, John T. *Understanding Torture: Law, Violence, and Political Identity*. Ann Arbor: University of Michigan Press, 2010.

Peters, Jeremy W. 'Pentagon Reinstates Banned Guantánamo Reporter'. *New York Times*, 9 July 2010. Viewed on 29 May 2016. https://mediadecoder.blogs.nytimes.com/2010/07/09/pentagon-reinstates-banned-guantanamo-reporter/.

Public Committee Against Torture in Israel. *Prosecutorial Indifference: Systematic Failures in the Investigation of Soldier Violence Against Detainees in the Occupied Palestinian Territories*. Jerusalem, June 2014. Viewed on 29 May 2016. http://stop-torture.org.il/wp-content/uploads/2015/10/%D7%91%D7%9E%D7%A6%D7%97-%D7%9C%D7%90-%D7%A0%D7%97%D7%95%D7%A9%D7%94-%D7%92%D7%A8%D7%A1%D7%94-%D7%91%D7%90%D7%A0%D7%92%D7%9C%D7%99%D7%AAProsecutorial-Indifference.pdf.

'UK Was Involved in Libyan Torture Flights and Politicians Knew, Say British Prosecutors'. *Reprieve*, 9 June 2016. Viewed on 10 June 2016. http://www.reprieve.org.uk/press/uk-was-involved-in-libyan-torture-flights-and-politicians-knew-say-british-prosecutors/.

Rejali, Darius. *Torture and Democracy*. Princeton and Oxford: Princeton University Press, 2007.

'Rendition: Blair in Quotes'. *BBC*, 19 January 2006. Viewed on 29 May 2016. http://news.bbc.co.uk/2/hi/uk_news/politics/4627360.stm.

Roach, Kent. *The 9/11 Effect: Comparative Counter-Terrorism*. New York: Cambridge University Press, 2011.

Rosenberg, Carol. 'Alleged 9/11 Plotter's Lawyers Ask Prosecutor, Judge to Quit Trial over Destruction of Evidence'. *Miami Herald*, 11 May 2016. Viewed on 29 May 2016. http://www.miamiherald.com/news/nation-world/world/americas/guantanamo/article77015207.html.

Sadat, Leila. 'Ghost Prisoners and Black Sites: Extraordinary Rendition under International Law',. *Case Western Reserve Journal of International Law* 37.2 (2006): 309–342.

Save the Children (Sweden). *One Day in Prison – Feels Like a Year: Palestinian Children Tell Their Own Stories*. Stockholm, 2003. Viewed on 29 May 2016. http://resourcecentre.savethechildren.se/sites/default/files/documents/2443.pdf.

Sheizaf, Noam. 'Beitunia Killings and the Media's Incredibly High Bar for Palestinian Stories', +972 *Magazine*, May 21 2014. Viewed on 29 May 2016. http://972mag.com/beitunia-killings-and-the-medias-incredibly-high-bar-for-palestinian-stories/91166/.

Tagg, John. *The Burden of Representation: Essays on Photographies and Histories*. Minneapolis: University of Minnesota Press, 1988.

Terdiman, Richard. 'Translator's Introduction to Pierre Bourdieu's The Force of Law: Toward a Sociology of the Juridical Field', *Hastings Law Journal* 38 (1987): 805–813.

U.S. Congress, Senate, Select Committee on Intelligence, *Committee Study of the Central Intelligence Agency's Detention and Interrogation Program*. 2014. Viewed on 29 May 2016. https://www.amnestyusa.org/pdfs/sscistudy1.pdf.

Viterbo, Hedi. 'Seeing Torture Anew: A Transnational Reconceptualization of State Torture and Visual Evidence', *Stanford Journal of International Law* 50.2 (2014): 281–317.

Whitty, Noel. 'Soldier Photography of Detainee Abuse in Iraq: Digital Technology, Human Rights and the Death of Baha Mousa', *Human Rights Law Review* 10.4 (2010): 689–714.

Winter, Yves. 'Violence and Visibility', *New Political Science* 34.2 (2012): 195–202.

Wolff, Robert Paul. 'On Violence', *Journal of Philosophy* 66.19 (1969): 601–616.

Žižek, Slavoj. *Violence: Six Sideways Reflections*. New York: Picador, 2008.

CHAPTER 3

Facing Evil: Can We Professionally Evaluate Torture?

David Senesh

Abstract

Torture is a case of man-made relational trauma with devastating mental consequences for the victims that undermines basic beliefs about world order and human nature. This is equally true for mental health professionals who have borne witness to torture or who treat torture victims, as proximity to evil can trigger intense reactions and bear grave consequences for lay and professional witnesses alike, especially for those who document and evaluate cases of extreme maltreatment and abuse. Therefore, it is imperative that we delineate therapists' range of possible responses. In search of an optimum range of personal equanimity and professional integrity, there is a risk of setting rules of thumb and standardization based on an elusive sense of experience and familiarity. There is an apparent paradox when experts' gained knowledge and experience in documenting and assessing torture may lead to professional distancing from the experience and from the victim who becomes part of a faceless crowd. As will be demonstrated in this chapter, the dialectical tension between accumulating experience and an attitude of 'not knowing' is critically important when dealing with experiences that are unrecognizable and unspeakable.

Keywords

torture – relational trauma – vicarious trauma – PTSD – witnessing

Introduction

Torture is often considered an extreme case of man-made relational trauma with devastating mental consequences for the victims that undermine basic beliefs about world order and human nature. The practice of torture is widespread, yet forbidden, by most national legislations and international treaties. Torture is applied worldwide for various purposes: to enhance interrogations;

 | DOI:10.1163/9789004408012_005

to induce terror, fear and shame; and to take revenge. When it is sponsored by the state it enjoys secrecy and provides impunity to its perpetrators. When seeking protection, help, justice, and compensation, victims of torture often face disbelief, denial, counter accusations and even threats, all of which may compound the trauma by adding alienation and social discredit to their mental devastation. Making matters worse, due to social apprehension and fear of terrorism, especially in the wake of 9/11, a number of mental health professionals in the United States and elsewhere harnessed their knowledge and skills in the service of national security, crossing the ethical boundaries of their own professional associations, such as the American Psychological Association (APA). When victims are subjected to abuse by torturers camouflaged as medical and/or mental health personnel, such transgressions may negatively impact the capacity of those subjects to trust clinicians if, and when, they later seek therapy, which undermines their rehabilitation. Without trust, it is difficult to establish the therapeutic rapport necessary for a curative relationship. Initially, the lending of psychological expertise to support the stratagems of national security gained the APA's blessing, embracing "malignant normality", in reference to Lifton's "normalization of evil" in the medical professions.[1,2] It was much later that such conduct created an uproar and incurred massive criticism.[3,4] As a result, therapists must manage the issue of torture in ways that are absolutely non-contingent on their personal attitudes or their feelings toward the backgrounds of clients. Naturally, this is extremely challenging when dealing with political opponents, criminals and terrorists (including 'ticking bombs') who complain of maltreatment by authorities.

Proximity to evil can trigger intense reactions and bear grave consequences for lay and professional witnesses alike, especially for those who document and evaluate cases of extreme maltreatment and abuse. Therefore, it is imperative that we delineate therapists' range of possible responses. In this chapter, I suggest a continuum that ranges from blind complicity with perpetrating authorities on the one hand, to complete identification with clients on the other, both pertaining to the loss of critical thinking. In search of an

---

1 R. J. Lifton, *The Nazi Doctors*. (New York: Basic Books, 1986).

2 R.J. Lifton, *The Broken Connection*. (Washington: American Psychiatric Press, 1979).

3 K. S. Pope and Thomas G. Gutheil, "Contrasting ethical policies of physicians and psychologists concerning interrogation of detainees," *British Medical Journal*, 338, (2009), 1863. K. S. Pope and Thomas G. Gutheil. Psychologists abandon the Nuremberg ethic: Concerns for detainee interrogations. *International Journal of Law & Psychiatry* 32, no. 4 (2009),161–166.

4 K. S. Pope and Melba J. T. Vasquez, *Ethics in Psychotherapy and Counseling: A Practical guide (5th ed)*, (Hoboken: Wiley, 2016).

optimum range of personal equanimity and professional integrity, there is a risk of setting rules of thumb and standardization based on an elusive sense of experience and familiarity. When experts defend themselves from secondary vicarious traumatization by feeling more 'at home' and at ease, they fail to be in touch with the hovering weirdness that is the essence of the aversive experience. There is an apparent paradox when experts' gained knowledge and experience in documenting and assessing torture may lead to professional distancing from the experience and from the victim who becomes part of a faceless crowd. As will be demonstrated later in this chapter, the dialectical tension between a sense of being 'experienced' and an attitude of a novice of 'not knowing', is critically important when dealing with experiences that are unrecognizable and unspeakable. Paraphrasing Thomas Ogden, each encounter with every act of torture described by a victim/survivor should be met with a fresh professional reaction of awe, without prior experience and with no expectations.[5]

Professional Positioning and Attitudes towards Torture

For those who brave these difficulties, there are other threats to their professional standing and integrity. The accounts by tortured individuals might be biased, and/or punctured with missing data and contradictions, due to dissociation, forgetfulness and repression. These are all integral characteristics of the post traumatic condition of stress. When interviewing the victim of torture, assessors are called on to re-construct and re-structure the trajectories of the victim's story as an integrative and clear whole, and account for the discrepancies. They must fill in the gaps and apply their skills to help embed and validate such claims of tortures with the added value of coherence, reliability and validity.[6] The employment of the standard evaluation procedures of the Istanbul Protocol[7] allows for post factum physical and mental symptoms to connote the application of specific methods of torture. The familiarity and experience of experts with the wide array of techniques and tactics employed

5 Thomas H. Ogden, *The Primitive Edge of Experience*. London: Jason & Aronso Books, 1989.

6 David Senesh, "Facing Evil: Can we evaluate the psychological consequences of torture" (in Hebrew) *Psycho-actualia* April 2015, 29–32.

7 UN Office of the High Commissioner for Human Rights (OHCHR), *Manual on the Effective Investigation and Documentation of Torture and Other Cruel, Inhuman or Degrading Treatment or Punishment ("Istanbul Protocol")*, 2004, HR/P/PT/8/Rev.1, available at: http://www.refworld.org/docid/4638aca62.html [accessed 22 September 2018].

in torture provides a professional advantage compared with knowing about and advocating against torture solely on the basis of the existing literature, combined with a solid stand on human rights. A collaborative and practical investigation by legal, medical and mental health professionals provides a powerful mechanism that approximates this objective, helping to describe what may otherwise remain indescribable, and to illuminate what otherwise may remain obscure.

The Special Features of Relational Trauma in Torture

Man-induced traumas form an especially severe category of mental injury often accompanied by moral consequences. In contrast, natural disasters and accidents, including most forms of physical and medical discomfort, lack the psychic components of intention and volition, not to mention the resulting sense of regret, shame and guilt. Moreover, since they occur in the interpersonal arena they violate the infrastructure of human attachment and trust formed during infancy. Such trauma is often referred to as 'intimate trauma' or 'betrayal trauma.' Sometimes this concept is extended to include cases when nature 'betrays' us, or when our body 'fails' us, and we tend to personalize such occurrences. When people betray us, we transcend the boundaries of humanity and apply such terms as 'inhuman', 'bestial', 'monstrous', and so forth. Such discourse fails to recognize the boundaries of what counts as human (if not humane) behavior. Following Holmes and Rahe,[8] the typical exclusion of torture from lists of major life events and traumas is illustrative not just of our inattention regarding torture, but also our intolerance for its inclusion in the repertoire of human life. Such inattention and intolerance is effectively an abdication of responsibility for torture's eradication.

Unlike other definitions of psychological disorders, the application of forensic and medical categories to define trauma as a disorder requires external validation. The two cases most relevant to torture are those traumas that occur in man-induced violence (such as aggressive criminal action and war) and those that occur to people who are dependent and helpless – children, the mentally handicapped, the incarcerated, the marginalized and otherwise subordinated. When torture is applied within what are considered privileged relationships, they constitute an additional extreme case of intentional transgression.

8 Thomas Holmes and Richard Rahe, "Stressful Life Events." *Journal of Psychosomatic Research* Vol 11, (1967), 213.

Therefore, both Reactive Attachment Disorders (RAD) and Post Traumatic Stress Disorder (PTSD) are anchored on assumed occurrence of external events and transactions and not contingent merely on perceptions or the individual's intrinsic predispositions and vulnerabilities. They are also considered to have pervasive, malignant and resistant lifelong consequences to the individual and to those around him. The usual DSM category of post-trauma thus refers to experiences that cannot be contained and processed by the human psyche, nor by her social milieu. Thus, another concept of complex-PTSD has evolved and been put forward to test such cases when the trauma is inter-subjective and leads to devastating results in subsequent relationships. On top of the physical, psychological and social impacts, there are consequent ruptures in the post-traumatic world-view of order, fairness and morality and the fusion of the self within such a world. The resulting sense of detachment and alienation may result in a chronic sense of 'existential homelessness.'[9]

Reaching the Core Experience of Torture Induced Trauma

The core experience of torture appears ever evasive, resistant to mentalization let alone verbalization and communication to others. Language alone fails to describe the experience. Physical, artistic, ecological and other therapeutic means are harnessed in order to reach beyond what words can convey. Without these means, the proximity to evil blankets the subject, placing his experiences far into the domain of the unthinkable and unspeakable.

Preceding Amery's 'existential homelessness,' there were significant attempts to describe that sense of alienation following such extraordinary experiences. First and foremost, it was Freud in 1919 who termed the 'unheimlich' as something that inhabits the region beyond the limits of one's familiar perceptions and knowledge. The 'unheimlich' casts a conscious sense of threat that is nonetheless obscured by remaining partly in the unconscious. It is the uncanny, or, oxymoronically, 'the unknown known'. This is apparently as 'close to home' as we can get when trying to describe the inconceivable.

Frustration and isolation are not experienced exclusively by the victims. Attempts to capture the essence of torture-related traumatic experiences and squeeze them into medical categories result in formal and dry terms which are unsatisfactory and frustrating for professionals. The attempts to medicalize and legalize the experiences induced by torture run the risk of banality

9 Jean Amery. *At the Mind's Limits: Contemplations by a survivor on Auschwitz and its realities.* (Bloomington: Indiana University Press, 1980), 21–41.

and technicality. They are stripped of the nearness of experience and instead form abstract categories of violations, symptoms and damages. Professionals are struggling to define, measure and quantify such experiences and their consequences without the loss of their lived impact. They try to balance evidence-based approaches with experience-near methods of assessment and present them in detailed case studies and case presentations. The need for tangible evidence and empathy-provoking descriptions all too often entails graphic descriptions, detailed accounts, and sometimes precise firsthand citations. The need to quantify the functional consequences of trauma by the application of psychological tests and measures is balanced against methods that are benevolent and sensitive to survivors' fragility and risk of re-traumatization during the interview. Procedural interviewing methods that do not adhere flexibly to these vulnerabilities may result in harming the victims due to the interviewers' empathic fatigue and burnout. On the other hand, overly empathic and compassionate witnessing may put the professional witness at risk of over-identification and secondary vicarious traumatization, and consequently pose the additional challenge of defending his seemingly biased conclusions in a formal report or in court. Either way, there are major risks in attending to the toxic stories of abuse and torture because they attack the basic common linkages and profound tenets of our worldview and sense of existence. As a result, they may induce feelings of powerlessness, disgust, fear, shame and guilt. The witnessing professional often risks parallel professional and personal transformations which resonate with the victim's own horrible experiences.

The application of legal-medical-psychological methods of assessment to torture survivors, according to the Istanbul Protocol, seems to provide a balanced measure of precision and grace. The Protocol attempts to document violations of human rights and assess their physical and psychological consequences while not compromising their informants' wellbeing. Yet the application of the Protocol, without the proper training of professional agents, may not suffice in yielding the expected phenomenological precision that captures the core, or 'suchness', of torture. However, it may still be sufficient and satisfactory in providing the victim with a constructive narrative template for the purposes of treatment, rehabilitation and compensation. In trying to even the score without revenge, even the partial restoration of justice is a significant factor in the rehabilitation of faith in oneself and others, and thus may serve as a cornerstone in the process of healing.[10]

---

10 David Senesh, "Emerging Sense of Continuity When Facing Death and Trauma," in *Studies of Grief and Bereavement,* ed. Shulamith Kreitler and Henya Shanun-Klein (New York: Nova, 2013), 53–66.

In applying such assessment methods, there is an ample attempt to find a probable commensurability between the physical and psychological symptoms of torture and the victim's narrative. Personal stories of the violation of human rights can then be put within the larger historical and political context for additional verification. In addition to such tangible results, a sensitive and benevolent witnessing constitutes another therapeutic mechanism in and of itself, helping to restore a sense of humanity where it was so fatally compromised.

Obstacles and Pitfalls in the Assessment Process

The major obstacle to such endeavors is the typical refusal of society at large to admit that people are being subjected to state-approved torture. It is impossible for most people to hear and incorporate such horrendous truths while, at the same time, holding onto values such as freedom, dignity, equality and security. In short, such testimony contradicts what society holds sacred. In order to allow torture to continue in the interest of national security and provide impunity to the people designated to practice it, there is a schism between public discourse and the effectively subversive testimonies evidencing torture. Disavowal in the forms of secrecy and silencing is common practice to allow contradictory truths to exist. Such a response further damages the victim's psyche by provoking responses of denial, dissociation and repression. As public discourse is subdued or muted, so too is psychological processing.

So the process of witnessing, evaluating and documenting torture is required not only to expose and point out such major violations of human rights, but also to shed light on the practice of disavowal that allows them to prevail. Personal and collective defense mechanisms are put to work collaboratively to discredit such accounts, minimize the consequences, and disperse responsibility. Hence, it becomes paramount that professional ethics and personal courage challenge these mechanisms and barriers in order to help these precious and precarious accounts prevail through their documentation.

Baranowsky, Moses and Sundri[11] listed the challenges that clinicians face when working with victims of torture, as well as with victims of other human rights violations. Using a consensual, qualitative approach, they found that the difficulties hindering the practice include: lack of multicultural competence;

11 Kim A. Baranowsky, Melissa H. Moses, and Jasmine Sundri, "Supporting Asylum Seekers: Clinician Experiences of Documenting Human Rights Violations Through Forensic Psychological Evaluation," *Journal of Traumatic Stress* 31 no. 3 (2018): 391–400.

mastering trauma informed clinical interviewing procedures; isolation and lack of collaboration with other professionals; limited resources and lack of monetary compensation for time and work of the volunteers (i.e. a reliance on pro bono work). The most substantial of all the difficulties, however, is the transferal issues of the interviewer's own family and personal experiences. The sharing of unspeakable horrors can be seen as a double-edged sword to be handled with utmost care and respect. The ethics of listening requires total dedication in attending to the witness's history, taking account of all major life events and traumas in addition to incidences of abuse and torture. Both witness and victim walk the fine line between formal defensive discourse and empathic sharing in order to sufficiently protect themselves while at the same time arriving at healing.[12] In a recent work, Boulanger suggested that despite its damaging impact on professionals, vicarious trauma 'can be essential in helping both clinicians and patients process the traumatic material,'[13] provided that the contagious contents are dealt with within the framework of sustainable and necessarily elaborate therapeutic relationships. In cases of isolated, limited or sporadic forensic assessments, it is important to provide substantial supervision and therapeutic aftercare to both parties.[14]

Attempts to Formalize and Standardize the Assessment of Torture and Its Sequela

The Istanbul Protocol is one attempt to form a productive collaboration between multi-national and inter-disciplinary teams willing to courageously enter the abyss of torture. In doing so, they synergize the best of all professions and mobilize mutual support and combined resources to thoroughly work with each and every survivor of torture. Even when flooded with accounts of torture by many detainees or waves of incoming refugees, each account should be honored and attended to individually without presumptions and without resignation to routine and habituation. In working with a forensic group of Israeli and Palestinian legal, medical, and mental health professionals drawn from Israel and in the Occupied Palestinian Territories (OPT),[15] we operate in

12 David Senesh. "Trauma and Resilience: From Trench to Bench," *Traumatology* 23, no. 1 (2017): 18–22.

13 Ghislaine Boulanger, "When is Vicarious Trauma a Necessary Therapeutic Tool?" *Psychoanalytic Psychology* 35, no. 1 (2018): 60–69.

14 Ibid.

15 A commonly used term for Palestine.

the midst of an ongoing intractable political and military conflict. Trained to use the Istanbul Protocol, the team specializes in the assessment of mental health consequences of human rights violations among detained Palestinians and incarcerated refugees. The team itself consists of Israeli Jews and Israeli and OPT Palestinians, who might be divided on most political and social issues, yet insist on common moral and ethical grounds when caring for torture victims.

The members of the team collaborate despite their disagreement regarding the anteceding events that lead to the arrests, interrogations and torture of victims. The dialectic tension between the attributes of the victims and their actions (terrorism/freedom fighting, criminality) hovers over the entire process and requires meticulous attention, professional precision and collegial support. The team is continuously called to deal professionally with such concerns and contradictions. The denunciation of torture, as commonly defined by international treaties, sometimes serves as the only compass that keeps the team on track when dealing with its laborious, almost impossible, work.

An especially challenging incident occurred at one of our supervision sessions. Sometimes interviewers are carried away into excessive sympathy and over identification with the victim and lose their professional impartiality. However, sometimes identifications may split and polarize. A significant rift was revealed between physicians and mental health professionals when performing an assessment of a young, male Palestinian terrorist/freedom fighter who complained of torture while being interrogated by the Israeli General Security Services (GSS).[16] The physician experienced no moral dilemma while performing the physical check-up, but the Israeli-Jewish psychologist had difficulty conducing the amnestic interview and performing the Mental Status Exam (MSE).[17] In fact, even as a participant in the supervision session, I felt a moderate sense of secondary trauma with dissociative and avoidance reactions. When the psychologist admitted that it was almost impossible for her to conduct an empathic and compassionate act of witnessing of the torture victim, who theoretically could have formerly harmed her or other fellow Israelis, I was flooded by feelings of fear, anger and disgust which interfered with my sense of empathy and inability to remain emotionally attuned to the group. In discussing the matter further, it was suggested that in order to secure impartiality and neutrality in similarly charged cases, it might be advisable to conduct the interview and supervision by third party nationals.

16 The GSS is a covert agency that operates in Israel and the OPT.

17 MSEs evaluate an examinee's level of mental functioning, including time and place orientation, reality testing, judgment, perception, and memory.

Another incident manifested disagreement within the evaluating group of multinational professionals who differed on their views of an asylum-seeking refugee who allegedly was a victim of torture before arriving to Israel. Again, the physical evaluations stirred no dissension; however, the mental assessments by various professionals failed to keep an integrated empathic view of the victim of torture, particularly when describing predisposing drives and intentions to his violent and even criminal behaviors. These could be attributed to contextual factors, or to pre-morbid personality pathologies, such as immature, impulsive or psychopathic disorders. Different assessments produced different empathic positions among the experts, resulting in tension and animosity. Time and again, the team had to discipline itself to de-contextualize its attitudes by connecting to its basic ethical anchor that held torture as unacceptable under any circumstances, irrespective of prior conditions, behaviors and justifications. This was especially difficult when the victim of torture had victimized others or put others in his family or community at risk. These challenges require ongoing group support and personal supervision.

Secondary and vicarious traumatization in dealing with victims of tortures may result in either affective numbness, on the one hand, or over-identification and loss of critical perspective, on the other. Both seem to hamper the proper perspective required for professional work. Advocacy for human rights may sometimes blur clinical judgment and may result in discrediting the profession and its reputation. Furthermore, the application of overly lenient criteria (e.g. defining daily hassles as torture) may hamper the course of rehabilitation by providing unnecessary or overly generous secondary gains. In a deeper sense, the victim knows that their story was not really met with true empathy and authenticity, and thus their experience remains untouched, unaccounted and invalidated.

Equanimity in Evaluating Terror

Developing the knowledge base and competence in dealing with the phenomena of torture should not induce a false sense of familiarity and comfort. The sense of the unfamiliar and uncanny must be retained to offer victims a feeling of being understood and to give them a sense of being at home away from home. At the same time, the witness/interviewer should allow for a greater empathic curiosity to develop through benevolent attention and trust in the process of inquiry, which may constitute a journey of learning and specialization in what is a relatively new field, and the development of a more formal practice. Trauma and torture are always to be seen as meaningful and talking about them involves generosity and courage from both parties in order to yield precious experiential accounts. These must never be referred to as 'contents', 'sessions' or 'cases,' but as preciously humane transactions of shared suffering. They are precious because they form a sanctuary of trust where compassion and hope can once again be cultivated.

Personal and Professional Gain from Engaging in the Evaluation of Torture

Are there any prospects for personal or professional gain in engaging with the evil of torture? The basic tenet is that proper evaluation of victims will serve as a significant building block in their physio-psychosocial rehabilitation and in restoring their human rights, thereby promoting an optimal quality of life after the trauma of torture. The very recognition of one's victim/survivor/hero status and the resumed sense of connectedness with others may in itself welcome the abused person back to society. Although traumatic experiences cannot be undone, in some cases, proper legal and clinical processing may unleash strength and coping capacities that lead to personal growth. Meanwhile the encounter with these psychic materials can be toxic or 'radioactive' for the forensic professionals. Resulting sequelae should be measured against the prospects for personal growth and professional development.

For both parties in the evaluative encounter, if it is not of transformative magnitude, victims do not turn into survivors, and survivors do not evolve into heroes.[18] For the professionals who bring with them their own narratives (often loaded with major life events including personal and family traumas), the encounter with shared subjectivities of suffering has the potential to promote growth on both personal and professional levels. The most rewarding eventuality is when both parties benefit from such encounters.[19]

Additionally, the status of the mental health professions may be strengthened in following their commitment to their own ethical codes. Although the multinational and interdisciplinary composition of evaluation teams may run the risk of conflicts and ruptures, and therefore interfere with the delicate art

18 Leanh Nguyen, "Psychoanalytic activism: Finding the human, staying human." *Psychoanalytic Psychology* 29, no. 3 (2012): 308–317.

19 Ghislaine Boulanger, "Psychoanalytic witnessing: Professional obligation or moral imperative?" *Psychoanalytic Psychology* 29, no. 3. (2012): 318–324.

of evaluation, the encounter with such horror may set a different perspective, reframing such differences within a manageable scope. As victims become survivors, who then evolve into heroes, they may inspire their professional interlocutors to transcend their own differences. All parties then benefit from what is recognized as Post-Traumatic Personal Growth. This idea resonates in the form of two old Jewish sayings that 'in every bad there is some good' and that 'evil may breed good.'[20] Whatever appears as pure evil might pose a blessing in disguise. It might be argued cautiously that, given the proper trajectories of resources, training and attitudes, the horrific and totally unjustified evil of torture, when witnessed, documented and assessed, might breed some good and possibly beneficial post factum outcomes for the victims, the helping professions and to society at large.

Bibliography

Amery, Jean. *At the Mind's Limits: Contemplations by a survivor on Auschwitz and its realities.* Bloomington: Indiana University Press, 1980. 21–41.

Baranowsky, Kim A., Melissa H. Moses and Jasmine Sundri. "Supporting Asylum Seekers: Clinician Experiences of Documenting Human Rights Violations Through Forensic Psychological Evaluation," *Journal of Traumatic Stress* 31 no. 3 (2018): 391–400.

Boulanger, Ghislaine. "Psychoanalytic witnessing: Professional obligation or moral imperative?" *Psychoanalytic Psychology* 29, no. 3. (2012): 318–324.

Boulanger, Ghislaine. "When is Vicarious Trauma a Necessary Therapeutic Tool?" *Psychoanalytic Psychology* 35, no. 1 (2018); 60–69.

Holmes, Thomas and Richard Rahe, "Stressful Life Events." *Journal of Psychosomatic Research* Vol 11, (1967): 213.

Lifton, R. J. *The Nazi Doctors.* New York: Basic Books, 1986.

Lufton, R.J., *The Broken Connection.* (Washington: American Psychiatric Press, 1979).

Nguyen, Lenah. "Psychoanalytic activism: Finding the human, staying human." *Psychoanalytic Psychology* 29, no. 3 (2012): 308–317.

Ogden, Thomas H. *The Primitive Edge of Experience.* London: Jason & Aronso Books, 1989.

Pope, Kenneth S. and Thomas G. Gutheil. "Contrasting ethical policies of physicians and psychologists concerning interrogation of detainees," *British Medical Journal* 338, (2009), 1863.

20 In Hebrew "מעז יצא מתוק."

Pope, K. S. and Thomas G. Gutheil. Psychologists abandon the Nuremberg ethic: Concerns for detainee interrogations. *International Journal of Law & Psychiatry* 32, no. 4 (2009):161–166.

Pope, K. S. and Melba J. T. Vasquez, Ethics *in Psychotherapy and Counseling: A Practical guide (5th ed),* Hoboken: Wiley, 2016.

Senesh, David. A former prisoner of war as therapist. *Rahat Medical Journal* 2, no.1 (2004): 9–12.

Senesh, David. "Emerging Sense of Continuity When Facing Death and Trauma," in *Studies of Grief and Bereavement,* ed. Shulamith Kreitler and Henya Shanun-Klein. New York: Nova, 2013: 53–66.

Senesh, David. "Facing Evil: Can we evaluate the psychological consequences of torture" (in Hebrew) *Psycho-actualia,* April 2015; 29–32.

Senesh, David. "Trauma and Resilience: From Trench to Bench," *Traumatology* 23, no. 1 (2017): 18–22.

UN Office of the High Commissioner for Human Rights (OHCHR), *Manual on the Effective Investigation and Documentation of Torture and Other Cruel, Inhuman or Degrading Treatment or Punishment ("Istanbul Protocol")*, 2004, HR/P/PT/8/Rev.1, available at: http://www.refworld.org/docid/4638aca62.html [accessed 22 September 2018].

PART 2

Philosophical and Theological Perspectives

CHAPTER 4

The Ambiguity of Sovereignty: the Passion Narrative as a Paradox

Toni Koivulahti

Abstract

This chapter begins with Milbank's reflections on the torture of Christ. The aim is to examine alternative narratives that Milbank's thinking opens up regarding the universality of the experience of torture, from the standpoint of the Passion narrative. According to Milbank, the torture and death of Christ happened in a manner that resembles Agamben's writings on *homo sacer*, a person who exists outside the community and who can be killed without punishment, but not sacrificed in a religious ritual. For Milbank, who views Christ as structurally analogous to a *homo sacer*, the Passion is a ritual sacrifice and, therefore, a variant of the *homo sacer*. This opens up a new avenue for questioning Christ's execution by a way of examining his paradoxical position outside of the law, but at the same time 'crowned' the king of the Jews, and the son of God. This essay follows Derrida's three figures that exist outside the law: the beast, criminal, and sovereign. A theological understanding of these classifications can broach new avenues in analysing secular questions of torture and exteriority to the legal order. Christ's suffering combines these Derridean figures and ritualizes extra-legal violence, helping us understand how ritual violence becomes political violence.

Keywords

sovereignty – violence – narrative – Christianity ritual – sacrifice – torture

Introduction

Since its beginning, Christianity has been plagued by questions of the torture and execution of Christ, the reasons behind it, and inquiries into what actually transpired. The postmodern theologian, John Milbank, has attempted to answer these questions by examining Christ as a political actor who was reduced

 | DOI:10.1163/9789004408012_006

to 'bare life.' This idea of bare life, a life not protected or maintained by the law, has been intensively studied by European political philosophers. Interesting developments around this idea have been made around Giorgio Agamben's concept of the *homo sacer*. Agamben's theory has been discussed by various philosophers and theologians, among them Milbank, and the late French philosopher Jacques Derrida. Both Derrida and Milbank accept a crucial part of Agamben's analysis: the idea of a life placed outside the law, or a bare, animal life, which is the polar opposite of sovereignty, that is to say, of the supreme political power.

For Milbank, the idea of bare life is connected to the figure of Christ, and it is therefore crucial to ask what it means to say that Christ was, at his end, in a position of a bare, almost animal life. What would this animality mean when applied to – from the viewpoint of Christians – a God? Following Milbank, I will consider how the crucifixion could reconcile humanity by fleshing out the logic of political power. I will also consider whether, as Derrida has pointed out, the common silence of both God and animals has meaning in the context of Milbank's idea of Christ as a *homo sacer* and whether the idea of the *homo sacer* aids us in thinking of fresh ways to approach animal life as well as political constructions.

The Sacred Man

Giorgio Agamben's book *Homo Sacer: Il potere sovrano e la vita nuda* (1995), made a historical survey of the terrain of the biopolitical production of sovereignty, that is to say, how the uses of power relations have functioned to create political categories of humans that relate to their bodies. Central to Agamben's thesis of sovereignty is a figure from an obscure Roman law called *homo sacer*, a concept that is found throughout history under various names, but always as something that has been placed outside society, but also as something constitutive of itself. Agamben also argues that sovereignty is, at its core, both interior and exterior to the political order and the legal structures of the political system.[1]

For Agamben sacrality, or the concept of "the sacred," is something that precedes religion. He states that the sacred is the basis of any society of men. *Sacer* is something that is placed outside society that makes possible a

1 Giorgio Agamben, *Homo Sacer: Sovereign Power and Bare Life*, trans. Daniel Heller-Rozen (Stanford: Stanford University Press 1998), 71.

distinction between what is interior and what is exterior.[2] To avoid any ambiguity, Agamben states that for him the sacred is a mere semantic trace, not a reference to the conceptions of sacred that pertain to religious traditions.[3] Sacrality that concerns individual humans functions in a similar way. Agamben here relies on the Roman grammarian Pompeius Festus, who wrote that a sacred man, or a *homo sacer* is someone who has been convicted of a crime and can be killed without punishment but not sacrificed in a religious manner.[4] This exclusion from the sphere of the law and the freedom to kill an individual should not, according to Agamben, be considered a death penalty. He claims that the most ancient forms of capital punishment are in fact purification rites and therefore within the religious sphere – unlike the *homo sacer*, which is a political concept, not a religious one.[5] Agamben writes that the sacred exclusion that forms a society can be, and is, extended to individual persons. He claims that the basic, fundamental categorical pair in Western politics is the division between bare life and political existence.[6] That is, between those humans who can be considered 'sacred', which here means excluded from society, and those who have a political existence or that are included in the corpus of the society.

According to Carl Schmitt's famous dictum, 'the Sovereign is the one who decides on the state of exception'[7] Agamben claims that the ultimate power in society is, in a way similar to the sacred man, both outside and inside the political order. If the sovereign decides on the exception, the sovereign is both inside and outside of the law at the same time.[8] The sovereign is both the ultimate guarantor of law and order, and able to make exceptive decisions that are extralegal. The sovereign decision disrupts the legal order, but at the same time it is an effect of legal power.

Sacred, bare, life and sovereignty are therefore in strange way both inside and outside of the law. The *homo sacer* has been cast outside, but the whole societal structure hinges upon the fact that it can exclude persons when deemed necessary by the sovereign. A life that has no political existence, that is, one that has been deemed *sacer*, is structurally absolutely necessary to the political

2 Giorgio Agamben, *Language and Death: The Place of Negativity*, trans. Karen E. Pinkus and Michael Hardt (Minneapolis: The University of Minnesota Press 1991), 106.

3 Agamben, *Homo Sacer*, 112–115.

4 Ibid., 71.

5 Ibid., 81.

6 Ibid., 8.

7 Carl Schmitt, *Political Theology: Four Chapters on the Concept of Sovereignty*, trans. George Schwab (Chicago: The Chicago University Press 2005), 5.

8 Agamben, *Homo Sacer*, 15.

order, without it there would be no outside at all. It is therefore an integral part of society in that it facilitates the outside itself.[9]

The Passion Narrative

In the book *Being Reconciled: Ontology and Pardon* (2003), John Milbank contrasts the persecution and execution of Christ with Agamben's theory of the *homo sacer*. In a short, but remarkable, piece, Milbank asks a crucial question: How can we be sure that the *homini sacer* were not offered as a sacrifice? Milbank argues that a *homo sacer* could be killed without ritual purification, but this does not mean that his death could not be construed as a sacrifice. Milbank argues that here the matter is purely a question of degrees of ritual purification, and not of sacrifice itself. After this Milbank, displays his postmodern credentials fully by claiming: '[T]otal absence of ritual still belongs to ritual.' Milbank considers Jesus, as described in the New Testament, a sacrifice in the sense that is was not a punitive measure in response to an illegal act, but instead an effect of political interests, which he, contrary to Agamben, claims to have a historical potential to be sacrificial.[10] Milbank follows Agamben by claiming that political action of sovereignty is always within the realm of reducing the excluded life into 'quasi-animality', 'natural', or 'sub-human' status.[11]

Despite ominously leading the reader to the question, Milbank does not directly claim that Jesus was a *homo sacer* in the sense of being literally condemned under Festus' archaic law. He instead opts to claim that, although Jesus was not strictly a *homo sacer* in a straightforward sense of Roman legislation, he nonetheless can be shown to have been in a position that displays the double exclusion of both sovereignty and bare life.[12] So, while Jesus is not a *homo sacer* in the sense of being strictly in such a position within Roman law, he nonetheless belongs in the sense of Agamben's *homini sacri*. Milbank claims that Jesus was thrice within the structural place of a *homo sacer*: First, because he was abandoned by Jewish sovereignty to the Roman executive; second, when he was abandoned by the Roman executive to the mob of people; and, third, because (according to the Gospels of Luke and John) he was abandoned by the mob and given to Roman soldiers to execute.[13] Milbank

9 Ibid., 82.
10 John Milbank, *Being Reconciled: Ontology and Pardon* (London: Routledge, 2003), 90–92.
11 Ibid., 91.
12 Ibid.
13 Ibid., 92–93.

concludes that: '[T]he Cross exposed the structure of arbitrary sovereign power in its ultimate exceptional, yet typical, instance.'[14] In Milbank's deeply Christian view, this combination of a 'quasi-animal' creature, a historical political actor, and divine being, helps humanity to reconcile one person to the other in a universal manner.[15] That is, one aspect of this is that by displaying the violent functioning of society, this narrative can help us understand how ritual and political violence work.

Werewolves of Calvary

Aside from the religious content, this narrative displays how the violent workings of society have the capacity to expose the perverse workings of power. In this sense, one could further argue that to truly understand the politics of the passion of Christ one has to read it in a purely materialistic way. In this sense the passion could be read in a manner that the French philosopher Jacques Derrida calls 'fabular', that is to say a moral story that has a function, which is that it makes something known.[16]

What is interesting in reference to Derrida's thought here is that in his analysis of the problems of animality, bare life, and sovereignty, Derrida places his accent firmly on the question of bestiality. That is to say, for Derrida, the crucial question of exclusion is not limited to the 'quasi-animal' but to animality itself.

With regard to animality, Agamben has written of the figure of the bandit in ancient Germanic law. In this legislation, a criminal who could be killed in a similar way to the *homo sacer* was the *wargus,* or werewolf. The laws of Edward the Confessor, 1030–35, brought this half-animal figure into written law by calling the criminal who was banished to the outside of the law a *wulfesheud,* that is to say a wolf's head. This hybrid of a man and a beast is, again, a classic figure of the *homo sacer*. Divided between the city and nature it has no place to call its own.[17]

Milbank's focus on the structural details of the arrangement of sacrality, or banishment, demonstrates the importance he placed on the historical Jesus being a figure placed outside the juridical and political order, or even a *homo sacer.* This has to mean that, at least in this discourse, the figure of Christ is in a very profound

14 Ibid., 93.

15 Ibid, 103.

16 Jacques Derrida, *The Beast & the Sovereign: Volume I,* trans. Geoffrey Bennington (Chicago: The University of Chicago Press, 2009), 35–57.

17 Agamben, *Homo Sacer,* 104–105.

way linked to animality itself. In political terms, he is reduced from a political actor to a beast, from a man into a werewolf – a transformation from humanity to animality. Even more so, if one harmonizes Milbank's thesis with Jacques Derrida's observations of the relationship between bestiality and sovereignty.

Derrida sees man's proper place in the space between the beast and the sovereign.[18] He claims that a beast cannot be subject to the law, and therefore its place outside the law is in part due to the fact that it cannot transgress the law.[19] Derrida writes of the three figures: the sovereign, the beast, and the criminal and asserts that despite being obviously dissimilar, they share the feature of being in an extralegal position.[20] The beast, or wild animal, lives outside the juridical order; the criminal has abandoned it, and the sovereign exists above it.

If one contemplates Milbank's claims that the Crucifixion played out in a manner which corresponds to these postmodern concepts of the legal order, then Christ's position is inherently paradoxical. Christ would seem to be a criminal, a werewolf, but lacks even the political status of a condemned criminal. On the other hand, Milbank claims that the Crucifixion was a sacrifice, but a sacrifice of what precisely? The sacrifice of a God, clearly, but also of a purely animal humanity. This claim is actually not very controversial, but essentially consistent with the content of the so-called synoptic Gospels.

It would be foolish to misunderstand the content of the Gospels as reducible to simply the fable of an executed criminal. Within the Christian faith, Christ is God, therefore, as Paul wrote in his Epistle to the Romans, "So then he has mercy on whomever he wills, and he hardens whomever he wills."[21] Derrida has also pointed out, following the 17th century English political philosopher Thomas Hobbes, that the structure of sovereignty is originally a *mimesis* or modelled upon divinity.[22] Not only is Christ a man, a beast, and a werewolf, but, as a God figure, he is the ultimate sovereign as well.

Conclusion

Derrida sees both God and Beast as doubly excluded. Not only is God silent, but in a more profound way he is eternally silent. We can never count on the response of God or be assured of his acceptance. For Derrida this is the

---

18 Derrida, *The Beast & the Sovereign Vol. I*, 98.
19 Ibid., 102.
20 Ibid., 17.
21 Rom. 9:18 ESV.
22 Derrida, *The Beast & the Sovereign Vol. I*, 46–47.

definition of absolute sovereignty – the sovereign does not have to respond if it is his prerogative not to answer. He is not, therefore, responsible to anyone for his acts. This points towards the fact that the sovereign is above the law. He has no responsibility to answer questions, and therefore a right to be irresponsible.[23] Here Derrida follows Schmitt's dictum: The sovereign is he who decides on the exception. Derrida, however, goes much further than Schmitt. For him, this mute sovereign cannot but appear 'stupid' or to use the French word: *bête*. *Bête* interestingly means both stupid, or dumb, and a beast. Like the beast, the brute, the dumb animal who cannot speak, the sovereign is mute. This kind of sovereign carries with him death, which as Derrida reminds us was according to the philosopher Emmanuel Lévinas not nothingness, but nonresponsive.[24]

What therefore unites the Beast with the Sovereign, the outsider with God, is unresponsiveness. Here Christ, the paradoxical beast, criminal, God, and sovereign is silent, not responding, but still, according to Milbank, supports structures underpinning the political order known by his actions. As the writer and lay theologian Gilbert Keith Chesterton has pointed out, Christ famously cried out on the cross: "My God, my God, why have you forsaken me?"[25] It was a bizarre moment of God himself being an atheist.[26] In a pivotal moment in the Christian narrative, at the instant of his death, God then became unresponsive even to himself. For Christianity therefore, the silence of God has a meaning beyond the silence itself: in the paradoxical moment of silence, the mute stupidity of God, there lies Christ the tortured man, the werewolf, and so the sovereign has his final victory.

According to Milbank, the figure of Christ reconciles human to human through the exposition of the arbitrary working of sovereignty. Derrida's beastly apparitions make one question however if this approach, which restricts its range to humanity, is far too limited in its scope. Should we not seek also this same reconciliation from both the silence of the beasts and the gods, and not only question how we deal with each other, but also with these, our others, as well?

Bibliography

Agamben, Giorgio. *Language and Death: The Place of Negativity*. Translated by Karen E. Pinkus and Michael Hardt. Minneapolis: The Minnesota University Press, 1991.

23 Ibid., 57.
24 Ibid.
25 Matt. 27:46 ESV.
26 G.K. Chesterton, *Orthodoxy*. (London: The Bodley Head, 1927), 255.

Agamben, Giorgio. *Homo Sacer: Sovereign Power and Bare Life.* Translated by Daniel Heller-Rozen. Stanford: Stanford University Press, 1998.

Chesterton, G.K. *Orthodoxy*. London: The Bodley Head, 1927.

Derrida, Jacques. *The Beast & the Sovereign: Volume I.* Translated by Geoffrey Bennington. Chicago: The Chicago University Press, 2009.

Milbank, John. *Being Reconciled: Ontology and Pardon*. London: Routledge, 2003.

Schmitt, Carl. *Political Theology: Four Chapters on the Concept of Sovereignty*. Translated by George Schwab. Chicago: The Chicago University Press, 2005.

CHAPTER 5

Barbarians at the Gate: Reasserting a Natural Law Definition of Torture

Lon Olson

Abstract

The approach to 'redefining' torture utilized by the George W. Bush Justice Department after 9/11 was founded on a particularly erroneous legal interpretation which led to human rights abuses that should have been prevented. In addition to the horrific suffering inflicted upon both the guilty and the innocent, this redefinition of torture led many Americans, civilian and military, to commit acts that both compromised their own moral integrity and besmirched the moral reputation of the nation they were sworn to serve. In contrast, I argue for a definition that is grounded in principles that emphasize respect for the dignity of human life as well as rationality, of both the torture victim and the torturer. In light of this, the definition of torture for which I argue relies heavily on natural law principles that articulate the dignity of all human beings. The end result is a definition of torture that has the capacity to serve as a foundation upon which to build more fine-grained laws designed to curtail this destructive practice.

Keywords

torture – natural law – water boarding – Abu Ghraib – pain – psychology – moral injury – Guantanamo Bay – Bybee Memorandum

Introduction

In the months following the terrorist attacks of September 11, 2001 the sense among many in the U.S. government was that America was facing something new in the annals of armed conflict: an enemy that would stop at nothing to inflict mass civilian casualties, even at the cost of their own lives. Barbarians were at the gate, and drastic measures were required to repel them. As

 | DOI:10.1163/9789004408012_007

combat operations in Afghanistan progressed and more Al Qaeda terrorists were captured, the opinion among many in the Bush Administration was that the use of coercive or 'enhanced' interrogation techniques, such as those employed against trainees at U.S. military SERE schools, would accelerate the process of intelligence collection by 'motivating' detainees to reveal critical information regarding impending attacks.[1] However, as these techniques are unlawful under both international and domestic law, the Bush administration requested that the Justice Department's Office of Legal Counsel (OLC) provide a formal legal opinion as to how these interrogation techniques could be justified.

This legal opinion, issued by the OLC in August 2002, came to be known as the Bybee Memorandum, and it made some startling claims.[2] First, its authors concluded that the word torture should be used only to describe the infliction of the sort of 'severe' physical pain that typically accompanies death or organ failure.[3] Second, the Memorandum's authors argued that captured Al Qaeda terrorists did not fit any of the categories outlined in the Geneva Convention, which defines prisoner of war status. Consequently, it was an open question, according to the OLC lawyers, whether Al Qaeda affiliated detainees, as members of a stateless terrorist organization, merited the humane treatment afforded soldiers fighting on behalf of a recognized political entity. Finally, the authors claimed that international and domestic laws prohibiting the use of torture could be construed as an unconstitutional infringement of the President's war-making powers as Commander-in-Chief.[4] Although the Bush Administration eventually retracted the Memorandum, the spirit behind its authorship continued to infect U.S. policy regarding the treatment of enemy combatants, directly contributing to flagrant human rights abuses at

1 SERE (Survival, Evasion, Resistance and Escape) schools prepare American aircrews, special operations forces and other 'prone to capture' troops for the type of treatment they can expect when captured, particularly by nations who are non-signatories to the Geneva Conventions. The goal is to provide service members with the tools to survive and evade capture and, if captured, to persevere under torture and, if able, escape and make their way back to friendly lines.

2 The memorandum was the product of OLC attorneys and was signed Jay C. Bybee, then Assistant Attorney General for the Office of Legal Counsel in the Bush Department of Justice.

3 Jeremy Waldron, 'Torture and Positive Law', *Columbia Law Review* 105.6 (2005): 1707. The idea of relating severe pain to that involving death or organ failure was drawn from statutes governing medical treatment and was meant to help govern scarce resource allocation in health care administration.

4 Jay C. Bybee, 'Memorandum for Alberto R. Gonzales, Counsel of the President Re: Standards of Conduct for Interrogation Under 18 U.S.C. §§ 2340-2340A', viewed: 29 May 2016, http://nsarchive.gwu.edu/NSAEBB/NSAEBB127/02.08.01.pdf.

Guantanamo Bay, Abu Ghraib and other U.S. and foreign detention facilities throughout the world.[5]

While the Bybee Memorandum contains several controversial assertions, I will primarily focus on the definition of torture that the Memorandum's authors promote as well as the legal and philosophical perspectives that informed their work. I do this to call attention to the danger of adopting a legal definition of torture uninformed by the most basic insights into human nature and which fails to account for the irreparable harm the practice of torture, in all its various forms, inflicts upon those involved in the practice. I will then propose a definition of torture grounded in natural law principles that account for torture's fundamental violation of human dignity and the moral harm that torture inflicts upon the tortured and the torturer alike.

The Trappings of Legality

Particularly alarming about the Bybee Memorandum is the claim that the descriptive term 'torture' should be confined to the infliction of the sort of severe pain associated with death or organ failure.[6] Fortunately, this claim is relatively easy to refute. Anyone who has experienced a toothache knows that, while not deadly, it can be painful enough to make any thought not directly related to relieving the pain nearly impossible. The authors of the Memorandum also failed to acknowledge that psychological pain can be as debilitating as physical pain. In terms of psychological anguish, anyone suffering from a particular fear, whether of snakes, enclosed spaces, or some other phobia, can attest to the extreme anguish that such fears precipitate. Despite this, the Memorandum grants license to a wide range of coercive techniques, such as sleep deprivation, water boarding and white noise – to name only a few. Most of these techniques are not deadly; however, the mental anguish they cause is often extreme.[7] That the definition of torture could be circumscribed such that it would allow techniques such as water boarding runs counter to our most common moral intuitions. As one legal scholar notes, 'Tinkering with the definition

5 David W. Bowker, 'Unwise Counsel', *The Torture Debate in America,* ed. Karen J. Greenberg (Cambridge: Cambridge University Press, 2006), 193.

6 Bybee, 'Memorandum for Alberto R. Gonzales, Counsel of the President Re: Standards of Conduct for Interrogation Under 18 U.S.C. §§ 2340-2340A'.

7 Dave Grossman, *On Combat* (Warrior Science Publications, 2008), 23. Humans will die from sleep deprivation more quickly than they will from starvation.

of torture gives one the feeling that something sacred and inviolate is being tampered with.'[8]

The definition of torture defended in the Bybee Memorandum was largely motivated by the controversial belief that torture is a reliable means by which to motivate prisoners to part with closely held secrets.[9] Given the fear and anxiety that was prevalent after 9/11, fueled by the very real possibility of catastrophic follow-on attacks, the motivation to endorse coercive physical methods can be understood, if not condoned. While my main argument does not rest on the unreliability of the information extracted through torture, it bears mentioning that, despite prevailing opinion, the information gained through coercive methods is highly suspect. As Shane O' Mara has observed, the effects of torture-induced stress on the human brain are significant and severely hamper victims' ability to engage in 'meaningfully conscious directed recall of individual episodic memories.'[10] This is reinforced by field reports from several sources, which indicate that interrogators are routinely surprised at the lack of intelligence gathered using coercive methods. The scientific research of O' Mara and others indicates that rather than yielding an intelligence bonanza, the actionable information obtained through torture is so paltry that proponents of torturing detainees are left with an indefensible case when torture is compared with other effective, non-coercive methods.[11]

Aside from concerns regarding the reliability of intelligence extracted through torture, what is equally, if not more, disturbing is the absence of ethical insight that the OLC attorneys brought to their duty of providing legal advice to the U.S. President. Rather than providing their client, the President and his national security staff, with clearly defined legal boundaries, the OLC lawyers carved away at the international torture statutes until they had arrived at a definition of torture so exceptionally narrow that the Dean of the Yale Law School referred to it as 'perhaps the most clearly erroneous legal

8 Waldron, 'Torture and Positive Law', 1710.

9 Douglas A. Pryor, *The Fight for the High Ground* (Fort Leavenworth: CGSC Foundation Press, 2009), 6. 'Contrary to widely held belief, torture is not a highly effective method for extracting information from detainees. This assertion is supported by the experience of several veteran interrogators. Generally, the most effective methods for interrogation involve painstakingly building a relationship of trust with the prisoner over the course of weeks and months.'

10 Shane O'Mara, *Why Torture Doesn't Work: The Neuroscience of Interrogation* (Cambridge: Harvard University Press, 2015), 144.

11 Ibid., 2.

opinion I have ever read ... offering a definition of torture so narrow that it would have exculpated Saddam Hussein.'[12] In reading the Memorandum, what becomes apparent is that its authors are seeking to provide a patina of legality to a preordained conclusion, not a good faith legal analysis of international and domestic law pertaining to torture. In their authorship of the Memorandum, the OLC attorneys acted more like mafia *consiglieri* than as officers of the court with a sworn duty to interpret conscientiously both the letter and spirit of the law. Cloaking torture in the trappings of legality not only makes a mockery of the law, but also allows torture to spread and infect other segments of society. As Jeremy Waldron observes, '[t]orture has shown itself not to be the sort of thing that can be kept under rational control.'[13] This is borne out by experience. The coercive techniques authorized in the Bybee Memorandum were originally meant for the top echelon of Al Qaeda detainees, such as Khalid Sheikh Mohammed, the mastermind behind the 9/11 attacks. However as interrogators transferred from one theatre of operations to another, there was considerable cross-pollination of interrogation techniques. The coercive techniques employed against the Al Qaeda detainees at Guantanamo Bay soon migrated to Iraq where they were used on common Iraqi insurgents and even on innocent Iraqi civilians caught up in military dragnets. To make matters worse, some interrogators expanded on the methods authorized in the Memorandum and took it upon themselves to improvise even more abusive techniques. In one instance, a senior Baathist general was duct-taped into a sleeping bag, beaten, and eventually suffocated to death during an interrogation session.[14]

By failing categorically to prohibit coercive interrogation techniques, the Bush Administration created an environment that allowed many interrogators to surrender to what might best be described as 'the darker angels of their nature.' Taking this into account, it is important to define torture in such a way that not only expands the type of pain that qualifies as torture, thus sealing off avenues by which sophists can twist the definition to suit political ends, but that also recognizes the moral harm torture imposes on victim and torturer alike and how this harm is transposed to a society that condones it.

12 Michael Traynor, 'Citizenship in a Time of Repression', *Stetson Law Review* 35.3 (2006), 777.

13 Waldron, 'Torture and Positive Law', 1717.

14 Pryor, *The Fight for the High Ground*, 52–53.

Deriving a Definition from Natural Law Principles

The most articulate and inclusive definition of torture is provided by the United Nations International Convention Against Torture, referred to more simply as the Convention, or the CAT. This code, ratified by 77 nations in 1987, requires a nation to 'take effective legislative, administrative, judicial or other measures to prevent acts of torture in any territory under its jurisdiction.'[15] and defines torture in the following way:

> For the purposes of this Convention, the term "torture" means any act by which severe pain or suffering, whether physical or mental, is intentionally inflicted on a person for such purposes as obtaining from him or a third person information or a confession, punishing him for an act he or a third person has committed or is suspected of having committed, or intimidating or coercing him or a third person, or for any reason based on discrimination of any kind, when such pain or suffering is inflicted by or at the instigation of or with the consent or acquiescence of a public official or other person acting in an official capacity. It does not include pain or suffering arising only from, inherent in or incidental to, lawful sanctions.[16]

This definition encompasses a range of behaviors and motives that should rightly be proscribed. Unlike the definition of torture proposed in the Bybee Memorandum, the Convention definition encompasses the inimical effects inherent in the infliction of mental suffering. Further, it does not restrict the definition of 'severe' pain by confining it to some extraordinarily narrow standard. By specifically making reference to the sanction of public officials, the Convention captures a particular facet of torture – its institutional application – that makes the practice of torture so virulent. The Convention's definition of torture also encompasses a wider spectrum of situations in which torture is routinely used, apart from extracting information from uncooperative prisoners. In a move to preempt complaints that the physical and mental hardships associated with legal incarceration also meet the definition of torture, the Convention definition makes a crucial distinction between the physical and mental suffering that characterize torture from the discomfort that is inherent

15 *Convention Against Torture*, New York, 10 December 1984, United Nations Human Rights Office of the High Commissioner, available from http://www.ohchr.org/EN/ProfessionalInterest/Pages/CAT.aspx.

16 Ibid.

and unavoidable in the detention and punishment of legitimate lawbreakers. However, this is the same definition of torture that the authors of the Bybee Memorandum exploited to such devastating effect. As noted previously, the OLC lawyers twisted the descriptor 'severe' to justify a broad range of objectionable practices that clearly violate any common-sense conception of torture. In light of this, a stronger definition is required, one that firmly closes the door on this perceived loophole. Natural law theory is in a position to provide a more robust definition.

In its simplest expression, natural law theory endorses the view that there are four natural inclinations that promote human wellbeing: preservation of life; procreation; acquisition of knowledge; and forging social networks.[17] Actions that support these inclinations contribute to wellbeing while those that run counter to them undermine it. As with any philosophical theory that claims to assert fundamental truths of human nature, this view has been challenged. For instance, many critics note that natural law has been employed to advocate predominately patriarchical societies and institutions that are restrictive of equal rights for women, the LGBT community, and other minority groups. These critics justifiably worry that a natural law perspective can be harnessed to support an objectionably narrow view of what sort of choices are conducive to living a moral life. However, a more expansive interpretation of natural law theory maintains that these inclinations leave a great deal open to individual interpretation regarding of how they may be satisfied.[18] Natural law theory focuses on the value of human reason, maintaining that through reason we possess the capacity to discern the moral from the immoral, and to incorporate these determinations into the decisions we make regarding how to live flourishing lives. Natural law theory places a premium on human reason, viewing it not only as the source of human dignity, but also as essential to establishing certain moral obligations in terms of how we respect the dignity of others.

The practice of coercive interrogation is a clear violation of at least two of the inclinations referenced above: the inclination to preserve life (abstain from harm); and the inclination to forge and maintain social connections. To illustrate this, one must first come to terms with what the practice of torture is ultimately meant to accomplish. As Francesco Campagnoni observes, '[t]orture

17 C.E. Harris, *Applying Moral Theories* (Belmont: Wadsworth Publishing, 2007), 128–129.

18 Christine McKinnon, *Character, Virtue Theories and the Vices* (Ontario: Broadview Press, 1999), 22–23. McKinnon argues for a view of natural inclinations that supports the idea that there are a variety of ways to live an ethical human life and that there is no single best way to flourish as a human being.

tends to the disintegration and consequent annihilation of the psychic and moral personality, to the non-physical destruction, practically speaking, of the human person, with lasting result.'[19] Campagnoni identifies what lies at the heart of torture: the psychological destruction of the human psyche and the deliberate stripping away of the dignity that distinguishes the victim as a human in the first place. In light of this, the general goal of the torturer is to turn his victim into someone who is isolated, overwhelmed, terrorized, and humiliated.[20] Commonsensically, the natural law injunction to preserve life includes refraining from the sort of psychological annihilation described above. Perversely, the authors of the Bybee Memorandum endeavored to thwart established prohibitions against torture by advocating techniques, such as waterboarding and sleep deprivation, that limit, if not avoid, lasting physical evidence of abuse. However, even though physical evidence of torture may be minimized, psychological trauma is always evident, demonstrated most by the prevalence of severe post-traumatic stress (PTS) among survivors of torture – a type of psychological trauma that incidentally also extends to torturers.

In addition to violating the proscription to preserve life, torture violates the natural human need to form social ties. One of the most inimical aspects of the post-traumatic stress brought on by torture is that of 'moral injury' experienced by the victim. Moral injury was a concept familiar to the ancient Greeks, who saw it as a violation of the concept of *themis,* a term that has no exact English equivalent, but which can be roughly translated as 'right order or custom', and can be expanded to refer to the inherent order of the world or universe.[21] Anything that upsets this order, such as a physical attack or a gross violation of personal dignity, leads to severe psychological anguish, fatally undermining the victim's trust in humankind. This sense of moral violation and destruction of trust is experienced not only by victims but also by those who victimize others and, in the process, violate the moral standards with which they themselves were raised, particularly if the torturer receives official sanction for his crimes. Without this trust, positive social interaction is exceedingly difficult, if not impossible. Understandably, victims of torture often experience enormous problems with re-integrating into society once they return home, which in turn adversely effects their relationships with family and friends.

19 Cited in Edward Peters, *Torture* (Philadelphia: University of Pennsylvania Press, 1999), 187.

20 Zachary R. Calo, 'Torture, Necessity and Supreme Emergency: Law and Morality at the End of Law', *Valparaiso University Law Review* 43.4 (2009), 1592.

21 Jonathan Shay, *Achilles in Vietnam: Combat Trauma and the Undoing of Character* (New York: Schribner, 1994), 5.

As noted, the degradation brought about by torture can be accomplished by both physical and psychological means and is not contingent solely upon the severity of the pain inflicted. Indeed, the circumstances under which pain is experienced influence the victim's perception of the degree of that pain. As Edward Peters notes, '[t]he pain produced by torture is very likely to be stronger and perceived more intensely than a clinically comparable amount of pain might be if it occurred accidentally or in the course of disease etiology.'[22] Furthermore, given the differences in pain thresholds from one individual to the next, and varying psychological predispositions to certain fears, it is difficult to quantify exactly what level of physical or psychological suffering qualifies as torture, as any attempt is necessarily arbitrary. A stress position that might cause discomfort to a young, fit soldier may be excruciatingly painful to an elderly person suffering from arthritis or a degenerative bone or nerve disease. Some individuals find confinement within an enclosed space to be relatively stress-free, while others find it panic-inducing. In his excellent study of the history and practice of torture, Peters cites work done by Ronald Melzack and Patrick Wall who make the following observation in their book, *The Challenge of Pain*:

> Psychological evidence strongly supports the view of pain as a perceptual experience whose quality and intensity are influenced by the unique past history of the individual, by the meaning he gives to the pain-producing situation and by his 'state of mind' at the moment. We believe that all these factors play a role in determining the actual patterns of nerve impulses that ascend from the body to the brain and travel within the brain itself. In this way pain becomes a function of the whole individual, including his present thoughts and fears as well as his hopes for the future.[23]

Variations in the way different individuals experience and perceive pain dramatically expand the concept of torture to include methods and techniques that the authors of the Bybee Memorandum deliberately excluded from their definition, such as sensory and sleep deprivation, prolonged solitary confinement, and exploitation of specific phobias that the victim may possess that, while not painful in the traditional sense, cause debilitating psychological effects that undermine the will and dignity of the prisoner. All of these

22 Edward Peters, 166.

23 Ibid.

techniques have the capacity to produce significant personality disintegration that is just as damaging in the long term as overt physical abuse.[24]

Clearly, definitions of torture that set a high bar on the severity of pain can be exploited to allow methods that violate any commonsense understanding of torture. While the infliction of physical and mental suffering is integral to the practice of torture, the intensity of pain is not the only objectionable part of the practice. As philosophers William Twining and Barrie Paskins observe:

> The point has often been made that in many actual cases intensity of pain, whether physical or mental, is not the most objectionable element in the situation: what is typically treated as more serious is interference with the consciousness or will or dignity of the victim. It is a somewhat artificial extension of the notion of pain to include all these considerations, but it is equally artificial to treat techniques of interrogation, coercion and so-called "brain-washing" which do not involve crude physical violence or acute mental distress, as falling outside the present enquiry. Psychiatric treatment of political dissidents, sensory deprivation and other techniques for softening up, breaking the will or otherwise manipulating people, and solitary confinement may all fall outside of the traditional notion of torture.[25]

As described above, torture is as much about undermining the victim's dignity as it is about inflicting pain. In the infamous Abu Ghraib prison, Iraqi prisoners were routinely forced to strip and remain naked in the presence of female guards and pose for pictures which depicted them in degrading sexual situations; for a prisoner raised in the sexually conservative atmosphere characteristic of most Arab societies, appearing nude in front of a woman is the epitome of shame. Such shame and the resultant loss of face can lead many prisoners to attempt suicide.

Despite critical flaws in the definitions of torture cited above, there is no reason to rebuild a definition of torture from the ground up. Considering this, I propose making a small yet meaningful revision to the definition of torture set forth in the Convention:

---

24 Jason M. Breslow, 'What Does Solitary Confinement Do To Your Mind?' *Frontline* 22 April 2014, viewed 01 June 2016, http://www.pbs.org/wgbh/frontline/article/what-does-solitary-confinement-do-to-your-mind/.

25 William Twining and Barrie Paskins, 'Torture and Philosophy', *Proceedings of the Aristotelian Society* 52 (1978), 157.

> Pain, suffering or humiliation, whether physical or mental that is intentionally inflicted on a person for such purposes as obtaining from him or a third person information or a confession, punishing him for an act he or a third person has committed or is suspected of having committed or intimidating or coercing him or a third person, or for any reason based on discrimination of any kind, intentional debasement of a person's human dignity through the application of physical pain and/or mental suffering, for the purpose of subverting that person's free will or that of a third person.[26]

Whereas the original definition states that to qualify as torture the pain inflicted must be 'severe', the modified version I propose contains no test of severity. Instead, the intentional debasement of human dignity serves as the benchmark for whether an act qualifies as torture. In stripping away the victim's dignity, whether through physical and psychological pain or humiliation, the torturer seeks to control the will of the victim, which is the key to everything else the interrogator is seeking. Considering this, the inclusion of the infliction of humiliation to the definition is another key change. Many of the guards and interrogators at Abu Ghraib demonstrated a sadistic talent for exploiting Arab cultural sensitivities by inflicting various forms of sexual humiliation upon detainees, exacerbating their sense of lost dignity. Differing cultural sensitivities to certain types of humiliation, such as public nudity, illustrate the difficulty of tying a definition of torture to specific methods and techniques. The range of potentially humiliating practices is virtually limitless; it is unavoidable that something offensive will be left unaddressed. Therefore, the addition of humiliation is an improvement over previous definitions.

Finally, natural law also acknowledges an overlooked dimension of torture, which is the debasement of the torturer himself. There is a widespread assumption that torturers can be neatly categorized into two classes: either torturers are essentially sadistic by nature, drawn to the profession by their own dark impulses, or they are patriotic servants of the state performing an unpleasant yet necessary duty. In the latter case, many believe that these men and women possess the moral fiber that allows them to remain psychologically unaffected by their gruesome task. However, this view is contradicted by Paul Gibert, a veteran of the French counterinsurgency in Algeria. Gibert, who left

26 *Convention Against Torture*, New York, 10 December 1984, United Nations Human Rights Office of the High Commissioner, available from http://www.ohchr.org/EN/ProfessionalInterest/Pages/CAT.aspx.

the military to join the priesthood, had the following to say regarding torture's degrading influence upon those who practice it:

> But let us admit for a moment that it might be possible to justify torture for the 'noble motives': have they thought for one moment of the individual who does it, that is, of the man whom, whether he wishes or not, one is going to turn into a torturer? I have received enough confidences in Algeria and in France to know into what injuries, perhaps irreparable, torture can lead the human conscience. Many young men have 'taken up the game' and have thereby passed from mental health and stability into terrifying states of decay, from which some will probably never recover.[27]

Gibert's view, informed as it is by practical experience, is echoed by others. In his 1993 encyclical, *Veritatis Splendor*, Pope John Paul II condemned torture as intrinsically evil based on the axiom of humans having a dignity bestowed by their creator which it is categorically wrong to debase.[28] Other theologians and philosophers, writing within the natural law tradition have supported this argument:

> Torture is an immoral option not only because it denies dignity to fellow human beings, but because it saps the humanity from those who employ it. This psychological dynamic is reflected in the fact that such infamies do more harm to those who practice them than those who suffer from the injury.[29]

The moral harm to which the torturer exposes himself was demonstrated in the infamous Stanford Prison Experiment conducted in 1971. In this experiment, Stanford psychologist Dr. Phillip Zimbardo and a group of colleagues recruited a cohort of college-age men for the purpose of simulating a prison environment in order to observe what happens when you place supposedly 'good people' in situations that have the potential to comprise their moral behavior. Before participating in the experiment, the volunteers underwent screening in an effort to exclude individuals who exhibited sadistic or

---

27 Peters, 179.

28 Pope John Paul II, Veritatis Splendor #80, August 6, 1993:[http://www.vatican.va/edocs/ENG0222/__P8.HTM], Date accessed: 31 Dec 2011.

29 Pope Benedict XVI, "The Problem With Torture: Excerpts From Pope Benedict XVI's World Day Of Peace Message, 2006": [http://www.usccb.org/issues-and-action/human-life-and-dignity/torture/the-problem-with-torture.cfm] Date accessed: 31 Dec 2011.

psychopathic tendencies. The remaining volunteers were divided into guards and prisoners. However, they were provided little in the way of explicit, formal instructions about how they should conduct themselves or what, if any, rules they should follow. The results of the experiment were both unexpected and disturbing. Within a few days, the moral conduct of the participants fulfilling the role of guards deteriorated to such a degree that they began to inflict substantial physical and mental abuse on the 'prisoners'. This included forcing them to strip naked, perform degrading tasks, and hurling verbal abuse at those who challenged the guards' orders. The abuse was so pronounced that many of the prisoners began to exhibit personality disintegration, with some even showing signs of suicidal tendencies. Although meant to last two weeks, the mental and physical state of the prisoners and the escalating brutality of the guards forced Zimbardo to terminate the experiment after only six days.[30]

Critics point out several methodological weaknesses with the experiment, such as a lack of a control group or established measures designed to isolate variables and measure a specific outcome. To make matters worse, Zimbardo himself abandoned the objectivity required of him as a researcher by assuming the role of 'warden' of the simulated prison.[31] Nevertheless, one conclusion that is uncontroversial is that the possession of power over other human beings that is unrestrained by legal, much less moral, boundaries has a deleterious effect upon the moral reasoning of those who exercise such power. In fact, many of the volunteers who acted out the role of guards suffered significant remorse in the years after the experiment was shutdown. Another reasonable assertion is that individual perversions are not the root cause of torture. Rather it is the systematic, official sanction of torture that elicits the aberrant moral behavior. In short, official sanction leads the agents administering the torture to exhibit behavior they might never otherwise display.

The inimical influence generated by the official sanction of torture was illustrated during the Greek torture trials of 1975. In 1967, a cabal of Greek military officers launched a coup that overthrew the Greek government, ushering in a military dictatorship that lasted until 1974. During its reign of terror, the Greek military junta made extensive use of torture, often recruiting young, impressionable men from the countryside, employing them as interrogators and indoctrinating them in the use of torture techniques. During the subsequent

---

30 Phillip G. Zimbardo, *The Stanford Prison Experiment: A Simulation Study on the Psychology of Imprisonment*. Accessed December 30 2017 http://www.prisonexp.org.

31 Maria Konnikova, "The Real Lesson of the Stanford Prison Experiment" *The New Yorker*, June 12 2015. Accessed December 2017. https://www.newyorker.com/science/maria-konnikova/the-real-lesson-of-the-stanford-prison-experiment.

trials of these interrogators, a father of one of the young defendants addressed the court: 'We are a poor but decent family. ... and now I see [my son] in the dock as a torturer. I want to ask the court to examine how a boy whom everyone said was a "diamond" became a torturer. Who morally destroyed my home and my family?'[32]

The father's anguish represents another aspect of torture: the cost that it exacts on the families of those who practice it. Just as the torture victim must find a way to reintegrate into society, so too must the torturer. When reintegrating, these men and women often return to roles as school teachers, coaches and police officers, to name only a few professions in which moral character plays a prominent role. When doing so, torturers bring all of the associated psychological freight that they acquired as torturers with them. As humans, we possess one character and it must serve us both in our professional as well as in our private lives. We cannot remove our character and hang it up as we would a lab coat or a military uniform before donning a different character for our lives with our families. Considering this, it is naïve to believe that torturers' professional lives do not adversely impact their relationships with their spouses and children, not to mention how they interact with their fellow citizens in wider society. For this reason if no other a society that endorses the use of torture does so at its own peril.

Conclusion

The fears and animosities that inevitably accompany war present a temptation to cast aside laws developed through calm deliberation. Many of the detainees imprisoned during the war on terror are guilty of gruesome crimes. Some, like Khalid Sheik Mohammed, the mastermind of the 9/11 attacks and the murderer of journalist Daniel Pearl, openly boast of murdering innocent civilians. It is hard to engender sympathy for many of these prisoners. They seem to have forsaken their humanity, and pleas for their humane treatment often fall upon deaf ears. It is easy under these circumstances, however misguided, to disregard the impact of coercive interrogation techniques on those who intend to cause us harm. Drawing attention to the damage torture inflicts upon the torturer, our fellow citizen and integral component of our collective moral community, might prove more persuasive than appealing only to the humanity of the tortured.

32 Cited in Peters, 179.

Since the end of the Second World War, the United States and other liberal democracies have formed a bulwark against the spread of torture. Regrettably, the Bybee Memorandum significantly breached that defensive wall. Fortunately, the Memorandum was retracted and the breach has been repaired – at least for now. However, maintaining the prohibition against torture requires constant vigilance. As of this writing, the Oval Office is now occupied by an individual who, at various times, has openly declared his support for torture, making it clear that he endorses measures far more extreme than anything contained in the Bybee Memorandum. Disturbingly, this is a view that enjoys support from many sectors of the American public.[33] The question of torture is far from settled in the minds of many well-intentioned but ill-informed citizens, and this willingness to consider torture as a viable option is particularly dangerous when it comes to those public officials charged with developing policy.

Aleksandr Solzhenitsyn succinctly expressed our susceptibility to moral corruption when he wrote that [t]he line separating good and evil passes not through states, nor between classes, nor between political parties either, but right through every human heart.'[34] This illustrates the importance of defining torture in such a way that takes into account the risks posed to those tasked with administering torture and the moral injury that torture inflicts upon even the most well-intentioned people. When repelling the barbarians at the gate, we should not be blind to the danger from within our own defences. We have the capacity to destroy from within that which the terrorists are attempting to destroy from without. David Maxwell Fyfe succinctly articulated the attitude civilized society should take towards torture, observing that '[b]arbarism is never behind us, it is beneath us. It is our task to see that it does not come to the surface.'[35]

Bibliography

Abramsky, Sasha. 'Exactly What Kind of Torture Does Donald Trump Want to Use?' *The Nation,* 25 March 2016. Viewed 3 June 2016. http://www.thenation.com/article/exactly-what-kind-of-torture-does-donald-trump-want-to-use/.

33 Sasha Abramsky, 'Exactly What Kind of Torture Does Donald Trump Want to Use?' *The Nation,* 25 March 2016, viewed 3 June 2016, http://www.thenation.com/article/exactly-what-kind-of-torture-does-donald-trump-want-to-use/.

34 Alexander Solzhenitsyn, *The Gulag Archipelago 1918–1956* (New York: Harper and Rowe, 1975), 175.

35 Cited in Peters, *Torture,* 147.

Breslow, Jason M. 'What Does Solitary Confinement Do To Your Mind?' *Frontline.* 22 April 2014. Viewed 1 June 2016. http://www.pbs.org/wgbh/frontline/article/what-does-solitary-confinement-do-to-your-mind/.

Bowker, David W. 'Unwise Counsel'. *The Torture Debate in America,* edited by Karen J. Greenberg, 183–203. Cambridge: Cambridge University Press, 2006.

Bybee, Jay C. 'Memorandum for Alberto R. Gonzales, Counsel of the President Re: Standards of Conduct for Interrogation Under 18 U.S.C. §§ 2340-2340A'. Viewed 1 June 2016. http://nsarchive.gwu.edu/NSAEBB/NSAEBB127/02.08.01.pdf.

Calo, Zachary R. 'Torture, Necessity and Supreme Emergency: Law and Morality at the End of Law'. *Valparaiso University Law Review* 43.4 (2009): 1591–1612.

Greenberg, Karen J. *The Torture Debate in America,* Edited by Karen J. Greenberg. Cambridge: Cambridge University Press, 2006.

Grossman, Dave. *On Combat.* Warrior Science Publications, 2008.

Harris, C.E. *Applying Moral Theories.* Belmont: Wadsworth Publishing, 2007.

McKinnon, Christine. *Character, Virtue Theories and the Vices.* Ontario: Broadview Press, 1999.

O'Mara, Shane. *Why Torture Doesn't Work: The Neuroscience of Interrogation.* Cambridge: Harvard University Press, 2015.

Peters, Edward. *Torture.* Philadelphia: University of Pennsylvania Press, 1999.

Pryor, Douglas A. *The Fight for the High Ground.* Fort Leavenworth: CGSC Foundation Press, 2009.

Shay Jonathan. *Achilles in Vietnam: Combat Trauma and the Undoing of Character.* New York: Schribner, 1994.

Solzhenitsyn, Aleksandr I. *The Gulag Archipelago 1918–1956.* New York: Harper and Rowe, 1975.

Traynor, Michael. 'Citizenship in a Time of Repression'. *Stetson Law Review* 35.3 (2006): 777–809.

Twining, William and Paskins, Barrie. 'Torture and Philosophy'. *Proceedings of the Aristotelian Society,* 52 (1978): 143–194.

Waldron, Jeremy. 'Torture and Positive Law'. *Columbia Law Review* 105.6 (2005): 1681–1750.

CHAPTER 6

Beyond the State: Human and Animal Positions outside the Law

Noora Koivulahti

Abstract

This chapter addresses the connection between torture and a position outside the law, a relation that is examined from the point of view of three social theorists, Hannah Arendt, Giorgio Agamben, and Costas Douzinas. Arendt argues that the only effective protection of human rights are national rights established by states; Agamben develops the concept of *homo sacer*, meaning a person living beyond the reach and interest of the law and at the same time being especially vulnerable because of that position; and Douzinas criticizes the concept of human rights as legitimizing violence committed by western liberal democracies. The starting point of all three is that grounds for torture of human beings are borne out of, and even necessitated by, exclusion from the state. Drawing on the work of Gary Francione, I argue that the same logic is structurally analogous with regard to the position of non-human animals, which are outside the law specifically because they are not legally recognized as possessing any rights that would not be overruled by the rights of their owners. This chapter argues that denying legal personhood makes torture possible, and even encourages it, both with regard to humans and non-human animals.

Keywords

animals – state – legal personhood – extra-legality – torture – human rights

Rights as a Solution to the Problem of the Underdogs

Recent years have seen an increasing interest in animal rights in both academic and legal contexts. Advancing animal rights has brought about cooperation from several fields of study, including ethology, zoology, anthropology, and philosophy. One example is The Great Ape Project, which advocates for rights to life, protection of individual liberty, and prohibition of torture for

 | DOI:10.1163/9789004408012_008

great apes.[1] Similarly, the Nonhuman Rights Project (NhRP) endeavours to win certain species of animals the legal rights of bodily liberty (not being held in captivity, for example in laboratories or circuses), and bodily integrity (not being touched without either consent or one's best interests in mind).[2] One of NhRP's recent court cases sought to free two chimpanzees, Hercules and Leo, from the University of Louisiana. The chimpanzees were used in experiments that involved inserting fine-wire electrodes into their muscles and routinely subjecting them to general anaesthesia.[3]

These kinds of organisations have developed out of an attempt to intervene on behalf of animals that continue to be exploited despite the growing amount of studies attesting to the fact that their mental and cognitive capacities in many cases match those of humans.[4] Sadly, such knowledge has not translated into legally codified commitments for improved moral treatment of non-human animals. Primates and other animals are still used in experimentation; held captive in zoos for entertainment; and exposed to the destruction of their natural habitats. Besides opening up questions of whose rights should be protected and on what grounds, the issue also forces us to ask whether increased rights would even be able to protect animals. The latter is justified in light of the various critiques that have been made of human rights. Despite supposedly being protected by rights, humans can fall outside of the law and its protection. The current refugee crisis is an obvious example of this problem. In the midst of Europe today, there are hundreds of thousands of people lacking any enforcement of their supposed human rights.

I claim that we should consider if animal rights face similar problems exhibited by human rights. I will examine the human rights critiques of political theorists Hannah Arendt, Costas Douzinas, and Giorgio Agamben dealing with exclusion. Using the works of legal scholar Gary Francione and theorist

---

1 Paola Cavalieri and Peter Singer, eds., *The Great Ape Project* (New York: St. Martin's Press, 1996).

2 'What Kinds of Rights Are We Seeking for Nonhuman Animals?' *The Nonhuman Rights Project*, 23 March 2013, viewed 25 May 2016, http://www.nonhumanrightsproject.org/2013/03/23/what-kinds-of-rights-are-we-seeking-for-nonhuman-animals/.

3 Lauren Choplin, 'Nonhuman Rights Project Chimpanzee Clients Hercules and Leo to Be Sent to Sanctuary', *The Nonhuman Rights Project*, 3 May 2016, viewed 25 May 2016, http://www.nonhumanrightsproject.org/2016/05/03/nonhuman-rights-project-chimpanzee-clients-hercules-and-leo-to-be-sent-to-sanctuary/.

4 For example: Alexander F. Skutch, *The Minds of Birds* (Texas: Texas A. M. University Press, 1996); Marian Stamp Dawkins, *Through Our Eyes Only? The Search for Animal Consciousness* (Oxford and New York: Oxford University Press, 1998); Donald R. Griffin, *Animal Minds: Beyond Cognition to Consciousness* (Chicago: University of Chicago Press, 2001); Virginia Morell, *Animal Wise: The Thoughts and Emotions of Our Fellow Creatures* (New York: Crown, 2013).

of speciesism Joan Dunayer, I will then analyse how related problems are seen in the arguments for animal rights that use their similarity to humans as the basis for rights.

Human-Likeness as a Reason for Animal Rights

The attempts to give rights to animals can be seen as part of a wider societal realization that the interests of animals are not adequately protected by existing statutes. Animal welfare laws, such as they are, have existed for hundreds of years. The first known animal welfare law was issued in Ireland in 1635 prohibiting the pulling of wool off sheep.[5] Nonetheless, they have generally only been able to protect animals insofar as protection has not too greatly undermined human economic and scientific interests. The position of animals as primarily human property has left them very vulnerable against violent and often even torturous practices by humans. As realized by many animal ethicists, giving rights to animals per se is problematic or even impossible because they lack legal standing. Animal rights projects such as the NhRP have been founded in collaboration with animal ethicists like Peter Singer, Steven M. Wise, and Tom Regan, who argue that only legal personhood, which establishes rights, can guarantee the much needed protection of animals.[6] They have argued for the moral status of animals based on their similarity with humans in capacities such as language and rationality.

These projects, however good their cause, are not without their problems. Indeed, they have been criticized precisely for leaning on a 'similar minds' position. This very anthropocentric position means that only certain species judged intelligent by human standards, such as apes, dolphins, and elephants, are deemed to deserve protection, while less human-like animals, such as various amphibian and rodent species, are relegated to unimportance. Cognitive ethologist Marc Bekoff has pointed out that categorizing animals into groups like 'higher' and 'lower' is extremely problematic because it fails to account for the worlds of the animals themselves, and that loaded words like 'smart' and 'intelligent' are often misapplied. We don't have, for example, an objective

5 Richard D. Ryder, *Animal Revolution: Changing Attitudes Towards Speciesism* (Oxford and New York: Berk, 2000), 49.

6 For example: Cavalieri and Singer, *The Great Ape Project*; Steven M. Wise, *Rattling the Cage: Towards Legal Rights for Animals* (London: Profile Books, 2000); Tom Regan, *The Case for Animal Rights* (Berkeley and Los Angeles: University of California Press, 2004).

scale to tell us that monkeys are smarter than dogs.[7] This perspective also assumes, often without much evidence, that everything human-like is valuable because humans are valuable, and thus contrasts humanity as something good with animality as something lower and less valuable.

Laws Create Lawlessness

Arendt has written about the problem of human rights in her book *The Origins of Totalitarianism*, first published in 1951. According to Arendt, the original problem of human rights was linked to their supposed inalienability, which directed attention to an abstract human being without nationality or place. Arendt noted how the problem came to light when a growing number of people started to appear whose rights were not safeguarded despite living in nation-states in the middle of Europe. In the tumults after World War I, many people from different nations, particularly Belarus, Greece, Armenia, and Bulgaria, fled their original countries. It came to the fore that when humans lacked their own government, there was no institution that would protect them. Arendt's major realization was that a loss of national rights was identical to a loss of human rights.[8]

Historically, Arendt traced the connection between a position without rights and a lack of nationality back to colonial Africa, where the British, in particular, dominated local populations. The imperialists discerned that by leaving the political status of the dominated populations uncertain, it was easier to control them.[9] According to Arendt, the Nazis copied this tactic when they laid the foundation for the destruction of the German Jewry by stripping them of their citizenship.[10] For Arendt this was proof that natural human rights did not exist. Excluded from citizenship of a sovereign state, placed in an extra-legal position, outsiders are made vulnerable against the potentially dangerous caprice of the enfranchised.[11]

This has also been argued by Douzinas. Douzinas has pointed out that if human rights were entitlements granted people on account of their humanity

7 Marc Bekoff, 'Deep Ethology, Animal Rights, and The Great Ape/Animal Project: Resisting Speciesism and Expanding the Community of Equals', *The Animal Ethics Reader*, eds. Susan J. Armstrong and Richard G. Botzler (London and New York: Routledge, 2003), 119–120.

8 Hannah Arendt, *The Origins of Totalitarianism* (Orlando: Harcourt Inc., 1976), 291–292.

9 Simon Swift, *Hannah Arendt* (London and New York: Routledge, 2009), 108.

10 Ibid.

11 Ibid., 112.

alone, instead of membership to a narrower group, such as a nation or class, then those without the laws of a particular group to protect them, such as refugees, could still rely on the protection afforded them by their basic humanity. But, because these people have no or very few reliable rights, the human rights described by liberal philosophy arguably do not exist. Rather, all actual rights are attached to citizenship.[12] As we are well aware, it is not the case that citizenship is given out freely. Refugees have to give up their citizenship when attempting to flee oppressive governments only to find that most nation-states are not willing to accept them. This is what Arendt described as the curse inherent to all nation-states. By their formation, they produce thousands, even millions, of potential refugees.[13]

However unacceptable to a liberal sensibility, we cannot disregard the fact that the exclusion of some groups of people is inevitable even in the context of the West's present predominantly liberal politics, a fact incorporated by Agamben into his theory of sovereignty. Following German political theorist Carl Schmitt, Agamben writes that sovereign is the one who decides on the state of exception, when or where laws of the State no longer apply. The creation of the state of exception, where we find the refugee, not only determines the threshold between the inside and the outside of the law, but also subsequently makes juridical order possible.[14] Sovereignty relies on deciding on the line between the inside and outside, whereby the juridical order determines the limits of its own power.[15] Somewhat paradoxically, the existence of laws always presumes, and reinforces, the existence of lawlessness. For Agamben, law has its power precisely in the capacity to maintain itself in relation to an exteriority – the division between inside and outside.[16] National laws, for example, are always based on the exclusion of non-citizens.

Attempting to Convert Moral Rights into Legal Rights

We could say that people without national rights are in a similar position to animals, since they have lost their legal personhood. Even though they should

12 Costas Douzinas, *Human Rights and Empire: The Political Philosophy of Cosmopolitanism* (Abingdon: Routledge-Cavendish, 2007), 99.

13 Arendt, *The Origins of Totalitarianism*, 290.

14 Giorgio Agamben, *Homo Sacer: Sovereign Power and Bare Life*, trans. Daniel Heller-Rozen (Stanford: Stanford University Press, 1998), 19.

15 Ibid., 15.

16 Ibid., 18.

naturally be protected by basic rights, in practice they often fall outside any system that would be capable of realizing those rights. Arendt was keenly aware of this likeness when she wrote about the attempts to arrive at a new bill of human rights after the war:

> [t]he groups they formed, the declarations they issued, showed an uncanny similarity in language and composition to that of societies for the prevention of cruelty to animals. No statesman, no political figure of any importance could possibly take them seriously.[17]

The moral seriousness of its goal is likewise what the animal rights movement attempts to prove. Legislators and courts must take animal rights seriously, and do so by giving animals the status of legal persons. But since the power of law and, therefore, the possibility of rights are based on categories of inclusion and exclusion, we must always ask who or what is to be excluded? In the case of humans, the barrier is built, amongst other things, between nationality and non-nationality. Of course, the case for animals is being built on other kinds of categories. The projects of ethicists like Wise and Singer seek to lift certain animals like apes and dolphins above the threshold of legal personhood, but at the same time leave other animals outside of it. Critics like Bekoff have pointed out that the grounds for the special moral status of these 'higher animals' are very arbitrary, and therefore the barrier they attempt to build is hard to see as justified. Wise lists attributes such as complex minds, self-consciousness, and the construction of complex societies as attesting to the moral value of chimpanzees, and therefore their suitability to have legal personhood.[18]

But the fact remains that the attempt to tie rights to certain capacities is ethically and philosophically very problematic. For one thing, it is problematic to use certain higher capacities as a criterion for rights since there will always be some humans, for example infants and also adults with severe disabilities, who do not possess those capacities, as Dunayer points out.[19] Secondly, it is often not taken in to account that several other species of animals besides great apes and cetaceans have been proven capable of abstract thought. For example, the dancing of honey bees is as symbolic as language, used to communicate the

---

17 Arendt, *The Origins of Totalitarianism*, 292.

18 Wise, *Rattling the Cage: Towards Legal Rights for Animals*, 244.

19 Joan Dunayer, 'The Rights of Sentient Beings: Moving Beyond Old and New Speciesism', *The Politics of Species: Reshaping Our Relationships With Other Animals*, eds. Raymond Corbey and Annette Lanjouw (Cambridge: Cambridge University Press, 2013), 29.

location of food.[20] Looking for human characteristics and declaring them the criterion for rights is tantamount to the circular argument that humans are morally most valuable because they are human.

Francione is a noted proponent for animal rights based on sentience only, and discounts the view that some animals, like the great apes, should be held as more deserving of moral status than other animals. Francione holds that sentience, which he defines as being perceptually aware, is the only morally relevant characteristic, because it contains the idea of having an interest in continuing to live, and of death being understood as harm.[21] Following Francione, I argue that if we want to extend legal personhood to animals, sentience would be a better criterion than humanlike capacities, for the latter work in too exclusive and limiting a manner.

Conclusion

Advocating certain rights for specific animals creates categories of inclusion and exclusion. If we can't help perpetuating these categories because of the structure of the law, we should at least be very careful to rationalise on what grounds these distinctions are made. When the animal rights movement promotes juridical rights for specific animals it excludes a huge number of others who also face abuse by humans and who, on moral grounds, should be seen as deserving of the same protections. Of course, no proponents of rights can escape the excluding character of the law. Recalling Agamben, the logic of the sovereign is always based on exclusion of one or more groups. There are always to be bodies – human or animal – outside of the law, left unprotected. Perhaps, then, when we are seeking justice and trying to advocate for the legal status of those stuck outside the law, we might have to forge what Agamben calls 'new politics'.

As a final note, the case for Hercules and Leo had a happy ending. The NhRP issued a press release on the 3rd of May 2016, which stated that the chimpanzees were to be released to a sanctuary along with a further 220 members of their species kept in captivity by the University of Louisiana. This happy ending reminds us of the possibilities, and prospects, of recognizing the moral value and potential legal personhood of animals.

20 Gavin Sherman and Kirk P. Visscher, 'Honeybee Colonies Achieve Fitness Through Dancing', *Nature* 419.6910 (2002), 920–922.

21 Gary Francione and Robert Garner, *The Animal Rights Debate: Abolition or Regulation?* (New York: Columbia University Press, 2010), 19–20.

Bibliography

Agamben, Giorgio. *Homo Sacer: Sovereign Power and Bare Life.* Translated by Daniel Heller-Rozen. Stanford: Stanford University Press, 1998.

Arendt, Hannah. *The Origins of Totalitarianism.* Orlando: Harcourt Inc., 1976.

Bekoff, Marc. 'Deep Ethology, Animal Rights, and The Great Ape/Animal Project: Resisting Speciesism and Expanding the Community of Equals'. *The Animal Ethics Reader*, edited by Susan J. Armstrong and Richard G Botzler, 119–124. London and New York: Routledge, 2003.

Cavalieri, Paola and Peter Singer, eds. *The Great Ape Project.* New York: St. Martin's Press, 1996.

Dawkins, Marian Stamp. *Through Our Eyes Only? The Search for Animal Consciousness.* Oxford and New York: Oxford University Press, 1998.

Douzinas, Costas. *Human Rights and Empire: The Political Philosophy of Cosmopolitanism.* Abingdon: Routledge-Cavendish, 2007.

Dunayer, Joan. 'The Rights of Sentient Beings: Moving Beyond Old and New Speciesism'. *The Politics of Species: Reshaping Our Relationships With Other Animals,* edited by Raymond Corbey and Annette Lanjouw, 27–39. Cambridge: Cambridge University Press, 2013.

Francione, Gary and Garner, Robert. *The Animal Rights Debate: Abolition or Regulation?* New York: Columbia University Press, 2010.

Griffin, Donald R. *Animal Minds: Beyond Cognition to Consciousness.* Chicago: University of Chicago Press, 2001.

Morell, Virginia. *Animal Wise: The Thoughts and Emotions of Our Fellow Creatures.* New York: Crown, 2013.

Regan, Tom. *The Case for Animal Rights.* Berkeley and Los Angeles: University of California Press, 2004.

Ryder, Richard D. *Animal Revolution: Changing Attitudes Towards Speciesism.* Oxford and New York: Berk, 2000.

Sherman, Gavin and Visscher, Kirk P. 'Honeybee Colonies Achieve Fitness Through Dancing'. *Nature* 419.6910 (2002): 920–922.

Skutch, Alexander F. *The Minds of Birds.* Texas: Texas A. M. University Press, 1996.

Swift, Simon. *Hannah Arendt.* London and New York: Routledge, 2009.

Wise, Steven M. *Rattling the Cage. Towards Legal Rights for Animals.* London: Profile Books, 2000.

PART 3

Historical and Literary Perspectives

CHAPTER 7

Torture, Slavery, Civilisation and Human Rights in the United States, 1820–1860

W. Fitzhugh Brundage

Abstract

The association of torture with slavery was essential to the de-legitimization of the institution as well as the articulation of nascent notions of human rights in the United States. One of the most remarkable episodes followed a sensational scandal involving the Lalauries of New Orleans in 1934, which fuelled the campaign by American abolitionists asserting that torture was innate to slavery. This contention acquired particular resonance at a time when other forms of violent coercion were under scrutiny in the United States. The outrages committed by slave owners, slavery's critics protested, could not be reconciled with the principles of enlightened civilization that the American republic should embody. Slavery was a relic of barbarism that should no more be tolerated (or worse, protected) than other archaic practices such as absolutism, polygamy, or human sacrifice. To level the charge of torture against slaveholders was to associate them in the American imagination with the Spanish Inquisition, Old World autocrats, Oriental despots, and bloodthirsty savages.

Keywords

American slavery – Madame Delphine Lalaurie – torture – abolitionism – civilization – barbarism – human rights

The Fire at the Lalaurie Mansion in New Orleans, Louisiana

When fire broke out in their New Orleans mansion on April 10, 1834, Delphine and Louis Lalaurie struggled to save their possessions from the flames. They rescued furniture and other accoutrements of their elegant lifestyle but displayed no evident concern about their nine slaves. As the fire spread, neighbours and Good Samaritans volunteered their assistance. The Lalauries rebuffed appeals to extricate their slaves, berating would-be rescuers

 | DOI:10.1163/9789004408012_009

for meddling where they were not welcome. Spurred in part by longstanding rumours about the Lalauries' mistreatment of their slaves, members of the crowd pushed their way into the slave quarters, where they discovered horribly abused slaves.

In the kitchen, where the fire had apparently been intentionally started, a badly abused and elderly slave cook was found chained to the floor. In another room, the searchers found mutilated and emaciated slaves suspended by their necks. One slave was described as 'rotten with sores, and in them were found numbers of living creatures.'[1] After removing the gagged and manacled slaves, the rescuers incarcerated them in the city jail – a step which, ironically, appears to have been in large part for their protection.

News of the fire and the gruesome discovery of 'barbarous and fiendish atrocities' at the Lalaurie mansion spread quickly, fanning 'popular fury.' Later that day, a crowd surrounded the mansion, which had been saved from the fire, apparently intent on imposing popular justice on the Lalauries. Delphine escaped harm by fleeing in a carriage, thereafter making her way to Mobile and finally Paris. Her husband disappeared as well, leaving the mob to direct its fury against the mansion, which they gutted, scarring the walls with graffiti, and smashing its lavish furnishings and strewing them along Royal Street.[2]

The scandal of the Lalauries gripped the city in the days after the fire. Both decorum and the inadequacy of words discouraged newspapers from full disclosure of the horrors the Creole couple had inflicted on their slaves. The New Orleans Courier confessed that the scenes in the Lalaurie mansion 'inspired in us so much horror that even at the moment of writing this article we shudder from its effects.' City residents, however, betrayed few qualms about gawking at the spectacle of the ruined mansion or the Lalaurie's traumatized slaves. Thousands of spectators swarmed the jail to inspect an array of manacles, neck rings, and devices of torture collected from the mansion and 'to witness the unfortunate wretches who had escaped cruelties that would compare with those of a Domitian, a Nero, a Caligula.'[3]

The Lalaurie scandal garnered national and international attention, becoming a flash point in the intensifying debate over slavery.[4] Embedded within the national argument over slavery during the three decades before the Civil

---

1 'Shocking Barbarities,' *New Orleans Mercantile Advertiser* 11 April 1834.

2 The principal contemporary sources for the Lalaurie atrocities are: *New Orleans Courier* 10–11 April 1834; *New Orleans Bee* 11–12 April 1834; *New Orleans Mercantile Advertiser* 11 April 1834.

3 *New Orleans Courier* 10 April 1834; *New Orleans Bee* 12 April 1834.

4 For international reports, see *London Morning Post* 22 May 1834; *London Standard*, 22 May 22 1834; *London Morning Chronicle*, 23 May 23 1834; *Bristol Mercury*, 24 May 24 1834; *Bell's*

War was one of the most sustained debates in American history about torture, coercion, and obedience. Although the Lalaurie scandal was only one, albeit sensational, piece of evidence in the litany of slave masters' cruelties compiled by antebellum abolitionists, it provides an apt starting point to consider debates about slave torture in antebellum America. With mounting certainty, opponents of slavery asserted that torture was innate to slavery. This contention acquired particular resonance at a time when other forms of violent coercion were under scrutiny in the United States. The outrages committed by the Laluaries and other slave owners, slavery's critics protested, could not be reconciled with the principles of enlightened civilization that the American republic should embody. Slavery was a relic of barbarism that should no more be tolerated (or worse, protected) than other archaic practices such as absolutism, polygamy, or human sacrifice. To level the charge of torture against slaveholders was to associate them in the American imagination with the Spanish Inquisition, Old World autocrats, Oriental despots, and bloodthirsty savages.

2 Abolitionists Link Slavery and Torture

The Lalaurie atrocities, of course, were not the first incidence of cruelty to slaves to attract attention. Critics of slavery seized on the Lalaurie affair in part because, like the horrors of the slave trade, it seemed certain to evoke a sympathetic response. Abolitionists relied on appeals to both perfectionist strands in nineteenth-century American Protestantism and sentiment to soften the hearts of white Americans to the long-tolerated cruelty of slavery. Central to the culture of sensibility that informed antislavery tactics was the

---

Life in London and Sporting Chronicle 25 May 1834; *Berrow's Worcester Journal*, 29 May 1834; [Edinburgh] *Caledonian Mercury*, 29 May 29 1834. Among the many published accounts of the Lalaurie scandal, see E. Thomas, *A Concise View of the Slavery of the People of Colour in the United States; Exhibiting Some of the Most Affecting Cases of Cruel and Barbarous Treatment of Slaves ….* (Philadelphia: E. Thomas, 1834); Edward S. Abdy, *Journal of a Residence and Tour in the United States* (London: John Murray, 1835), 392–393; La Roy Sunderland *Antislavery Manual* (New York: S. W. Benedict, 1837), 75; Harriet Martineau, *Retrospect of Western Travel* (London: Saunders and Otley, 1838), 263–267; Theodore Dwight Weld, *American Slavery as It Is: Testimony of a Thousand Witnesses* (New York: American Anti-Slavery Society, 1839), 91; E. Bunner, *History of Louisiana* (New York: Harper and Brothers, 1842), 252–254; and Benjamin Lundy and Thomas Earle, *The Life, Travels, and Opinions of Benjamin Lundy; Including His Journeys to Texas and Mexico, with a Sketch of Contemporary Events, and a Notice of the Revolution in Hayti* (Philadelphia: William D. Parrish, 1847), 113.

presumption that the visualization of experience, especially physical suffering, was essential to fostering sympathy.

When abolitionists placed the details of the scandal in New Orleans before readers, they sought to heighten the immediacy of the pain and suffering of the enslaved and to diminish the seemingly unbridgeable distance that separated the Lalauries' manacled slaves from newspaper readers in Boston, New York, and elsewhere. Abolitionists knew that they had to make visible and immediate the horrors of slavery to white Americans outside the South who lived, according to Garrison, in 'exceeding ignorance' of the institution.[5] As the Haverhill Gazette conceded when it justified reprinting the Lalaurie story, the realities of the institution of slavery could not be 'described or fully known except by observation.'[6] As public spectacles of bodily torment became less common in the United States, the challenge for slavery's opponents was to make more conspicuous the physical violence of slavery.

The vicarious spectatorship that ran through the reporting of the Lalaurie scandal, and the antislavery campaign in general, placed a premium on the most lurid and horrific examples of slavery's evils. It was these examples that would animate moral imagination, establish the humanity of slaves, and energize the crusade to destroy the institution. Because 'Language is powerless and inadequate to give a proper conception' of 'details which seem too incredible for human belief,' one account of the Lalaurie atrocities averred, 'We shall not attempt it but leave it rather to the reader's imagination to picture what it was.' Nevertheless, most reports provided sufficiently vivid details to aid readers in visualizing the physical distress of the Lalauries' slaves and to elicit sympathy. Bodies were described 'in deplorable condition' and 'covered with bruises and wounds from severe floggings.' Especially graphic was the description of the condition of one slave who 'had a large hole in his head: his body, from head to foot, was scarred and filled with worms!' Also highlighted was the torment endured by a 'mulatto boy, who had been chained for five months, being fed daily with only a handful of meal, and receiving every morning the most cruel treatment.'[7]

Intended to arouse sympathy and provide opportunities to display the appropriate sentiments, the accounts of the Lalaurie atrocities tutored readers as to their proper emotional response. Had they been present as witnesses, they were told, they too would have recoiled in horror and then been flooded by sympathy. 'We saw one of the miserable beings,' one account reported, 'The

---

5 William Lloyd Garrison to Ebenezer Dole, 14 July 1830, in Walter M. Merrill, ed., *The Letters of William Lloyd Garrison* (Cambridge: Belknap Press of Harvard University Press, 1971), I, 104.

6 *Haverhill Gazette*, 10 May 1834.

7 *New Orleans Courier*, 10 April 10 1834; *New Orleans Bee*, 11–12 April 1834.

sight was so horrible that we could scarce look upon it … The most savage heart could not have witnessed the spectacle unmoved.' Many commentators explicitly prodded readers to engage in imaginative empathy like that which Adam Smith had described. The Liberator, for instance, asked readers to 'conceive the pangs which those poor slaves must have suffered before their deliverers came.' Garrison's message was clear: any reader who failed to be moved by the Lalaurie scandal 'must be lost past all redemption.'[8]

By mobilizing sympathy and emotion on behalf of their cause, abolitionists aimed to transform their readers and audiences into witnesses against slavery in a national court of public opinion. Employing the conventions of courtroom and church disciplinary proceedings, abolitionists marshalled vivid documentary evidence, especially first-hand testimonials and damning news accounts, to indict slavery and slaveholders. Theodore Dwight Weld was explicit about the task laid before readers of his book, *American Slavery As It Is:* 'Reader, you are empanelled as a juror to try a plain case and bring in an honest verdict … A plainer case never went to a jury. Look at it.'[9]

In myriad instances, far less lurid than the Lalaurie outrages, the deliberate infliction of pain, especially against defenceless humans and animals, provoked controversy because it flouted the purported refinement of the age. Modern liberal societies, above all the United States, required of citizens public virtue and the subordination of destructive impulses through rational self-control and inner moral constraints. English philosopher John Stuart Mill captured the tenor of the era when he observed that 'the spectacle, and even the very idea of pain, is kept more and more out of the sight of those classes who enjoy in their fullness the benefits of civilization.'[10] By that measure, a reluctance to inflict cruelty was a mark of civilization, while a disposition toward cruelty placed one beyond the pale of civilized humanity. To surrender to passion, to display insensitivity to suffering, and to cause intentional pain was to mock the code of civilized society.

Evolving understandings the physiology of pain also influenced ideas about the moral imperative to prevent intentional suffering. By the early nineteenth century, advances in medical theory and practice began to ameliorate the experience of pain and raised the possibility that pain need not be an unavoidable accompaniment to the human condition. Rather than a divinely inflicted punishment or an atonement for sin, pain was increasingly thought to arise

8 *Boston Liberator*, 3 May 1834; [*Hudson*] *Ohio Observer*, 15 May 1834.

9 Weld, *American Slavery as It Is*, 7.

10 John Stuart Mill, 'Civilization,' [1836] in *Essays on Politics and Society*. Ed. By John M. Robson (Toronto: University of Toronto Press, 1977): 130–1.

from human action or inaction. Pain no longer needed to be endured or ignored because human intervention now could either prevent it or ameliorate it whenever it arose. Once pain was understood as a physical, mechanical sensation without redemptive qualities, then humans had a moral obligation to free others of the conditions that produced it.[11]

Having activated the moral sense of Americans who opposed the sanguinary criminal punishments of yore, reformers voiced a growing concern about tolerating deliberate pain not just at the hands of the state but wherever it might lurk. They drew attention to harsh physical discipline within private families that destroyed familial harmony and socialized children into cruelty. In response antebellum reformers proposed new state powers to protect battered wives and children, pressed for revision of divorce laws to include an expanded definition of marital cruelty, and advocated against corporal punishment of children in schools. Even the entrenched tradition of corporal punishment within the military, especially the Navy, came under serious attack.[12]

To the extent that any single tragic event could, the Lalaurie atrocities revealed the 'unparalleled barbarity' of slavery wherever it existed, including in New Orleans, one of the nation's richest and most cosmopolitan cities. In the eyes of slavery's opponents, the innate rights that each human possessed had been flagrantly transgressed by the Lalauries. For William Lloyd Garrison and many other commentators, the Lalaurie case illuminated the perverse and virtually unlimited prerogatives that slave masters enjoyed.[13]

---

11 Margaret Nicola Abruzzo, *Polemical Pain: Slavery, Cruelty, and the Rise of Humanitarianism* (Baltimore: Johns Hopkins University Press, 2011), Chapter One; Karen Halttunen, 'Humanitarianism and the Pornography of Pain in Anglo-American Culture,' *American Historical Review* 100 (April 1995): 303–334.

12 Among the important works on antebellum humanitarian reform are Norman S. Fiering, 'Irresistible Compassion: An Aspect of Eighteenth-Century Sympathy and Humanitarianism,' *Journal of the History of Ideas* 37 (April-June 1976): 195–218; (Michel Foucault, *Discipline and Punish: The Birth of the Prison* (New York: Pantheon Books, 1977);), 104–169; Myra C. Glenn, *Campaigns Against Corporal Punishment: Prisoners, Sailors, Women, and Children in Antebellum America* (Albany: State University of New York Press, 1984); Michael Ignatieff, *A Just Measure of Pain: The Penitentiary in the Industrial Revolution, 1750–1850* (New York: Pantheon Books, 1978); Michael Meranze, *Laboratories of Virtue: Punishment, Revolution, and Authority in Philadelphia, 1760–1835* (Chapel Hill: Published for the Institute of Early American History and Culture Williamsburg Virginia by the University of North Carolina Press, 1996); Elizabeth H. Pleck, *Domestic Tyranny: The Making of American Social Policy Against Family Violence from Colonial Times to the Present* (Urbana: University of Illinois Press, 2004); and David J. Rothman, *The Discovery of the Asylum; Social Order and Disorder in the New Republic* (Boston: Little Brown, 1971).

13 *Boston Liberator* 3 May 1834. Nearly two decades later, the *Westminster Review* reached the same conclusion while reflecting on the implications of the slave master's

As Garrison and others emphasized, the Lalaurie affair underscored the extent to which the formal law of slavery enshrined the property rights of masters while severely circumscribing the appropriate powers of the state and courts. For Garrison, there were few more telling illustrations of the archaic and barbaric nature of slavery. 'The master-passion in the bosom of the slaveholder,' he explained, 'is not gain, but the possession of absolute power, unlimited sovereignty.'[14] The institution of slavery, Garrison and other fumed, endured because for two centuries American slave owners had been granted life and death power over their human chattel.[15]

James Henry Hammond, a prominent pro-slavery activist and statesman, boasted that slaveowners were comfortable applying 'plantation justice' to punish challenges to their authority, petty crimes, thefts, minor affrays, and other intimate forms of misbehaviour. Usually after an impromptu trial of sorts by the planter or his agent, the master or his overseer administered a whipping, beating, or some other form of physical punishment to the errant slave.

In the slave autobiographies that proliferated mid-century, accounts of such proceedings were so ubiquitous that, in the words of abolitionist Samuel E. Cornish, they would 'make the heart to bleed and the ear to burn.'[16] Page after page recalled the crack of the whip, the lashing of flesh, the tearing of skin, the streams of blood, and the shrieks of pain that punctuated the lives of slaves. Former slaves catalogued a staggering array of punishments that demonstrated that masters employed terror no less than pain to impose their will.[17]

The horrors of slave torture included the sexual exploitation of female slaves, which was of a piece with the indolence, arrogance, lack of self-restraint, and tyranny that characterized the slaveholder's conduct. In the words of another opponent of slavery, the events in New Orleans had exposed 'a system that places our fellow beings … to be ravished and tortured without the power of resistance by the fiendish lusts, caprice, and malevolence of the worst of men.'[18] Henry Ward Beecher, perhaps the most celebrated minister of the

---

prerogatives: 'The slave is 'subject to the will of his master, or mistress, be she a Madame Lalaurie or no'' (143).

14 *Boston Liberator*, 1 March 1850.

15 Andrew Fede, 'Legitimized Violent Slave Abuse in the American South, 1619–1865: A Case Study of Law and Social Change in Six Southern States,' *American Journal of Legal History* 29 (April 1985), 94. See also Marvin L. Kay and Lorin Lee Cary, *Slavery in North Carolina, 1748–1775* (Chapel Hill: University of North Carolina Press, 1995), 73–74.

16 Samuel E. Cornish, *New York Colored American*, 4 March 1837.

17 Lewis Clarke, *Narratives of the Sufferings of Lewis and Milton Clarke* (Boston: Bela Marsh, 1846), 17.

18 [*Hudson*] *Ohio Observer* 15 May 1834.

era, explicitly invoked the spectre of Oriental cruelty and licentiousness when he claimed that compared to the South 'a Turkish harem is a cradle of virgin purity.'[19] Garrison raged against the apathy of Americans who turned their eyes as 'more than a half a million of adult females without any protection for their chastity' were 'constantly prostituted by their owners.'[20] Former slaves were even more forthright about the pervasiveness of sexual violence endured by slaves. Solomon Northup recounted the heart-rending predicament of Patsey, 'an enslaved victim of lust and hate,' who was under constant threat of rape at the hands of her master and of punishment at the hands of her jealous mistress. Harriet Jacobs, who herself had spent years thwarting her owner's efforts to rape her, charged that slavery made 'the white fathers cruel and sensual, the sons violent and licentious' while contaminating the daughters and making the wives wretched.[21]

The attention devoted to the cruelties inflicted by the Lalauries and other owners on their slaves was an indictment of the society that nurtured and harboured such human monsters. Indeed, the vilification of Delphine Lalaurie (and comparative inattention to her husband's role in the atrocities) was an amplification of the critique of what a Massachusetts newspaper called 'the deadening effects of slavery on the sensitivities of our nature.' None of the accounts established that Mme. Lalaurie alone was responsible for the condition of the slaves in her household. But her involvement in them, even if only as an accessory, made a mockery of the contemporary idealization of women as acutely sensitive and innately nurturing. Evidently, she was neither. Instead, she was variously described as a 'fiend,' 'a demon in the shape of a woman,' 'a wretch too base to live in a civilized land,' and a 'miscreant.' Garrison even mocked the convention of referring to her by the formal title of 'Madame.' So contrary to conventional feminine behaviour was Delphine's apparent sadism that the Providence Patriot posited that a 'hideous transformation' was necessary to turn 'one of the softer sex into a monster of cruelty.'[22] That 'a demon in

19 Henry Ward Beecher, quoted in D. Bartlett, *Modern Agitators: or, Pen Portraits of Living American Reformers* (New York: C. M. Saxton, 1859), 20.

20 *Boston Liberator*, 30 July 1831.

21 Solomon Northup, *Twelve Years a Slave* (Auburn: Derby and Miller; Buffalo: Derby, Orton, and Mulligan; and London: Sampson Low, Son & Co., 1853), 189; Harriet Jacobs, *Incidents in the Life of a Slave Girl* (Boston: np, 1861), 81.

22 [*New London*] *Connecticut Gazette* 7 May 1834; *Pittsfield* [*MA*] *Sun* 8 May 1834; *Boston Liberator* 10 May 1834; *New York Journal of Commerce* 10 May 1834; *Haverhill* [*MA*] *Gazette* 10 May 1834; *Boston Liberator* 25 May 10, 1834; *Providence* [*RI*] *Patriot* 2 August 1834. See Kathy Frances Morlas, 'La Madame et la Mademoiselle: Creole Women in Louisiana, 1718–1865,' (MA thesis, Louisiana State University, 2005), esp. 111–126;

the shape of a woman' displayed a taste for sadistic violence was irrefutable evidence of the irrationality, depravity, and barbarism that prevailed under slavery.

The inevitable consequence of the absence of checks on the power and cruelty of slave masters had been exposed by the Lalaurie affair. Accustomed to resorting to violence to impose their will, slaveowners like the Lalauries were desensitized to suffering and pain. In *American Slavery As It Is*, Theodore Dwight Weld diagnosed cruelty as 'the spontaneous and uniform product of arbitrary power.'[23] Where slavery prevailed, he continued, 'cruelty has become a habit' and 'the greatest cruelties' were inflicted 'quite unconsciously. Perhaps with the Lalauries' outrages in mind, Weld regretted the likelihood that the heart became 'by habit, not only totally insensitive to certain forms of cruelty, but even to find chief amusements in such cruelties.' In this manner, slavery accustomed all those under its shadow to violence, inhibiting the humane sensibilities upon which rested modern civilization.

That American society not only tolerated but protected the unbounded, arbitrary, and overweening power of slave masters was intolerable for many abolitionists. What further evidence did Americans need that slavery truly was a 'relic of barbarity,' wholly incompatible with modern civilization? To ignore the glaring evidence of inhuman cruelty that prevailed on southern plantations was to acquiesce to the perpetuation of an institution that made a mockery of boasts about American progress and civilization.

Torture, Slavery and the Campaign to Abolish Subordination

Slavery's defenders (and subsequent commentators, well into the twentieth century) pointed to the abolitionists' fetishism of cruelty and suffering as evidence of their prurient fanaticism. But such mockery underestimated the scope of the abolitionist critique of the brutality of slavery. Together, black and white abolitionists punctured the notion that torture and callous cruelty were something done exclusively by primitive people. Relentlessly, abolitionists laid bare the sordid reality that people – Americans – who boasted of being benevolent stewards of the most advanced civilization in history carried out deeds of pitiless cruelty that elsewhere in the world would have been denounced as acts of savagery. This indictment transformed the nation, its institutions, and

Shirley Elizabeth Thompson, *Exiles at Home: The Struggle to Become American in Creole New Orleans* (Cambridge: Harvard University Press, 2009), esp. 164–209.

23 Weld, *American Slavery As It Is*, 117.

its culture from a harbinger of modern, progressive civilization into a citadel of barbarism and tyranny.

In their campaign against slavery and American hypocrisy, abolitionists proposed an expansive understanding of freedom. They retained a healthy suspicion of the threat that state power could pose to a citizen's freedom. But abolitionists also worried about unjust coercion in private relations that had previously been regulated, if at all, by community tradition. At a time when the language and concept of rights were still immature, abolitionists staked out a position that challenged not only the ownership of human beings but also time-honoured patriarchal privilege in American society. Instead, abolitionists advocated a spacious concept of universal and inviolate freedom that did not reserve one set of rights for the privileged and powerful – patriarchs – while assigning another to their dependents. Individual sovereignty, in simple terms, should not secure the right of some to exercise personal authority over others. Elizabeth Cady Stanton, who was both an abolitionist and pioneering suffragist, explained succinctly, that there can be no 'independence where there is subordination to the absolute will of another.'[24] Accordingly, abolitionists insisted on the sanctity of each person's right to self-ownership and freedom from coercion.[25] Abolitionists committed themselves to consigning to the dust heap of history those institutions, laws, and folkways that protected the exercise of direct, absolute sovereignty over others. Thereafter, once slavery and patriarchal tyranny were expunged from the nation, two pernicious and long-standing engines of cruelty and torture would be purged as well.

Bibliography

Abdy, Edward S. *Journal of a Residence and Tour in the United States*. London: John Murray, 1835.

Abruzzo, Margaret Nicola. *Polemical Pain: Slavery, Cruelty, and the Rise of Humanitarianism*. Baltimore: Johns Hopkins University Press, 2011.

Bartlett, D. *Modern Agitators: or, Pen Portraits of Living American Reformers*. New York: C. M. Saxton, 1859.

Bunner, E. *History of Louisiana*. New York: Harper and Brothers, 1842.

24 Elizabeth Cady Stanton, *Elizabeth Cady Stanton, Susan B. Anthony, Correspondence, Writings, Speeches* (New York: Schocken Books, 1981), 34–35.

25 *Declaration of Sentiments and Constitution of the American Anti-Slavery Convention* (New York: American Anti-Slavery Society, 1835).

Clarke, Lewis. *Narratives of the Sufferings of Lewis and Milton Clarke.* Boston: Bela Marsh, 1846.

DeLombard, Jeannine Marie. *Slavery on Trial: Law, Abolitionism, and Print Culture.* Chapel Hill: University of North Carolina Press, 2007.

Fede, Andrew. 'Legitimized Violent Slave Abuse in the American South, 1619–1865: A Case Study of Law and Social Change in Six Southern States.' *American Journal of Legal History* 29 (April 1985): 94–93–150.

Fiering, Norman S. 'Irresistible Compassion: An Aspect of Eighteenth-Century Sympathy and Humanitarianism.' *Journal of the History of Ideas* 37 (April-June 1976): 195–218.

Foucault, Michel. *Discipline and Punish: The Birth of the Prison.* New York: Pantheon Books, 1977.

[Garrison, William Lloyd,] *Declaration of Sentiments and Constitution of the American Anti-Slavery Convention* (New York: American Anti-Slavery Society, 1835).

Garrison, William Lloyd, *The Letters of William Lloyd Garrison.* Ed. Walter M. Merrill. Cambridge: Belknap Press of Harvard University Press, 1971.

Glenn, Myra C. *Campaigns Against Corporal Punishment: Prisoners, Sailors, Women, and Children in Antebellum America.* Albany: State University of New York Press, 1984.

Halttunen, Karen. 'Humanitarianism and the Pornography of Pain in Anglo-American Culture.' *American Historical Review* 100 (April 1995): 303–334.

Ignatieff, Michael. *A Just Measure of Pain: The Penitentiary in the Industrial Revolution, 1750–1850.* New York: Pantheon Books, 1978.

Jacobs, Harriet. *Incidents in the Life of a Slave Girl.* Boston: np, 1861.

Kay, Marvin L. Michael, and Lorin Lee Cary. *Slavery in North Carolina, 1748–1775.* Chapel Hill: University of North Carolina Press, 1995.

Lundy, Benjamin, and Thomas Earle. *The Life, Travels, and Opinions of Benjamin Lundy; Including His Journeys to Texas and Mexico, with a Sketch of Contemporary Events, and a Notice of the Revolution in Hayti.* Philadelphia: William D. Parrish, 1847.

Martineau, Harriet. *Retrospect of Western Travel.* London: Saunders and Otley, 1838.

Meranze, Michael. *Laboratories of Virtue: Punishment, Revolution, and Authority in Philadelphia, 1760–1835.* Chapel Hill: Published for the Institute of Early American History and Culture Williamsburg Virginia by the University of North Carolina Press, 1996.

Mill, John Stuart, 'Civilization.' *Essays on Politics and Society*, edited by John M. Robson. Toronto: University of Toronto Press, 1977 [1836].

Morlas, Kathy Frances. *La Madame et la Mademoiselle: Creole Women in Louisiana, 1718–1865.* MA thesis. Louisiana State University, 2005.

Northup, Solomon. *Twelve Years a Slave.* Auburn: Derby and Miller; Buffalo: Derby, Orton, and Mulligan; and London: Sampson Low, Son & Co., 1853.

Pearson, Susan J. *The Rights of the Defenseless: Protecting Animals and Children in Gilded Age America.* Chicago: University of Chicago Press, 2011.

Pleck, Elizabeth H. *Domestic Tyranny: The Making of American Social Policy Against Family Violence from Colonial Times to the Present*. Urbana: University of Illinois Press, 2004.

Rothman, David J. *The Discovery of the Asylum; Social Order and Disorder in the New Republic*. Boston: Little Brown, 1971.

Stanton, Elizabeth Cady. *Elizabeth Cady Stanton, Susan B. Anthony, Correspondence, Writings, Speeches*. New York: Schocken Books, 1981.

Sunderland, La Roy. *Anti-Slavery Manual: Containing a Collection of Facts and Arguments on American Slavery*. New York: S. W. Benedict, 1837.

Thomas, E. *A Concise View of the Slavery of the People of Colour in the United States; Exhibiting Some of the Most Affecting Cases of Cruel and Barbarous Treatment of Slaves*. Philadelphia: E. Thomas, 1834.

Thompson, Shirley Elizabeth. *Exiles at Home: The Struggle to Become American in Creole New Orleans*. Cambridge: Harvard University Press, 2009.

Weld, Theodore Dwight. *American Slavery as It Is: Testimony of a Thousand Witnesses*. New York: American Anti-Slavery Society, 1839.

CHAPTER 8

Torture and Anti-Partisan War: the Case of the Italian Social Republic, 1943–1945

Federico Ciavattone

Abstract

In Italy, between 1943 and 1945, in addition to the war between German and Allied armies, a civil war was being fought in which torture played a prominent role. In the centre and north, there were the occupying Germans; in the south, the Allied forces who were pushing up the peninsula. Alongside the Germans was Mussolini, who, having been freed on September 23, 1943, went about setting up a new Fascist state, the Italian Social Republic (ISR). In northern Italy, an armed resistance sprang up to oppose the German occupation and the new Fascist government. In this civil war, the biggest problem for the ISR was the partisan movement, and eradicating partisan resistance became the overriding priority. In this effort, torture was widely practiced by the ISR, both in improvised facilities and in structures specifically designed for` the task. After providing a general picture of torture by the ISR, this chapter focuses on the use of torture by special units of the Republican Army. Although declared apolitical, these units had a specific purpose: eradicating the partisan phenomenon by any means necessary.

Keywords

fascism – Italian Social Republic – anti-partisan warfare – counterinsurgency – World War II – resistance – insurgency – civil war – violence – anti-partisan special forces

The Italian Social Republic and the Torture

On September 23, 1943, Mussolini, who had been freed by the Germans a few weeks earlier, returned to power, forming the Italian Social Republic (ISR) in central northern Italy. In Italy's south, under Allied control, Victor

 | DOI:10.1163/9789004408012_010

Emmanuel III and Pietro Badoglio responded to the establishment of the ISR by creating the Southern Kingdom. Within the territories of the new Fascist state, the situation was very complex: there was a government headed by Mussolini; an occupying ally (the Germans); and an armed resistance movement that operated in an anti-Fascist, anti-Nazi capacity. The stage was set for conflict, and Italy was spiralling ever deeper into the throes of civil war.

The civil war fought in Italy between 1943 and 1945 had all the characteristics of a total war, in which no distinction was made, by either side, between soldiers and civilians. As a result, Italians who supported the Social Republic felt justified in their use of violence, as all those who did not choose the ISR, as well as those who took a wait-and-see attitude, were perceived as enemies. From the Fascist point of view, this perception legitimized the use of violence in its multiple forms, from round-ups and reprisals, to executions and burnings, as well as torture.

The ISR was characterized throughout its organization by widespread use of violence against the partisan movement, against civilians, and against all those who were perceived as potential enemies. The use of violence became a common element shared by all the various branches of the armed forces and the Fascist police: the army; the navy; the Republican National Guard; the Black Brigades; the auxiliary police forces; and the autonomous squads. Both the German authorities and their allies in the ISR tolerated the use of torture. For their part, the Fascist units perfected their interrogation techniques and set up special spaces, both in barracks and in requisitioned houses (known as *ville tristi*), to practice this form of violence.

In a context of total war and 'globalization' of the fight against the partisans, the use of torture was a prerogative that was exercised not only by units that had a strong political and ideological component, like the Black Brigades or the Republican National Guard. Indeed, torture was widely practised by all the units that made up the armed forces and the police units of the ISR. This included the National Republican Army, an entity which, in the post-war period, neo-Fascists and veterans attempted to present as apolitical and extraneous to the violence of the civil war.

In this chapter, the purpose of torture by the Republican Army is examined by focusing specifically on two special units created for the purpose of eliminating the Resistance: the Apennine Hunters and the Anti-Partisan Group. What emerges from this analysis is that torture was practiced in order to satisfy two unique needs of the ISR: collecting the necessary information for striking the partisan bands, and as a tool of psychological warfare designed both to dehumanize the enemy and to intimidate them. Torture was how the adversary would be destroyed, physically and morally.[1]

The Necessity of Constructing a Propaganda Image of the Enemy

Not all Italians were aware of what it meant to face an armed resistance movement. The ISR therefore found itself having to construct a negative image of the partisan to justify the ongoing struggle. Conversely, Mussolini had to extrapolate from military thought all those ideals of gallant, loyal warfare which Fascism had propagandized for twenty years.

This was a one-sided representation that had a very specific purpose: to encourage a widespread demonization of the partisan and a rejection of his humanity. The process was viewed as necessary because only in this way could the ISR succeed in isolating the figure in the minds of Italians and thus succeed in denationalizing him. The ISR's goal was to create a representation of an individual who was alien to Italian culture, driven only by the 'bloody creed' of Communist ideology and 'Jewish-Allied' propaganda.

The Army and the Total War

'Globalization' of the anti-partisan effort required the complete, coordinated involvement of all the Republican armed forces and the ISR police components. The counterinsurgency effort saw the participation of Air Force foot battalions, Decima MAS Marine battalions, Republican National Guard combat groups and armoured units, special police units like 'Ettore Muti' Autonomous Mobile Legion, and auxiliary police battalions. However, particular attention should be drawn to the role of the National Republican Army and the Republican Fascist Party. The backbone of the National Republican Army was the four divisions commanded by General Graziani: the 'Monterosa' 1st Alpine Division, the 'Littorio' 2nd Grenadier Division, the 'San Marco' 3rd Marines Division, and the 'Italia' 4th Bersaglieri Division.[2] In addition to its organic units,

1 Toni Rovatti, *Leoni Vegetariani. La violenza fascista durante la RSI* (Bologna: Clueb, 2011), 148–151.

2 The numbering of the divisions is progressive and depends on the order in which the Large Units returned to Italy. Giorgio Pisanò, *Gli ultimi in Grigioverde* (Milan: FPM, 1982); Giuseppe Rocco, *Con l'onore per l'onore. L'organizzazione militare della RSI sul finire della Seconda Guerra Mondiale* (Milan: Greco & Greco, 1998).

the National Republican Army established two non-divisional groups to carry out counterinsurgency activities: the Special Units Training Centre (CARS), later named the Apennine Hunters Group, and the Anti-Partisan Group (RAP).

CARS was set up in Parma in March 1944. The objective of the Republican Army's General Staff was to create a training centre that specialized in modern insurgency and counterinsurgency techniques. Three regiments were formed under the name of the Apennine Hunters. During the instruction period, CARS made ample use of the principles and experiences of the anti-partisan war that had been pursued by Italy up to September 1943. However, the training mission was short-lived and in the summer of 1944, CARS went into a deep crisis. Created as a training centre, it was forced to neglect instruction because of the need to operate directly in anti-partisan operations.[3] The ISR Army General Staff decided to transform CARS into the operational Apennine Hunter Group, a light brigade-strength unit engaged in counterinsurgency operations. Initially, the Apennine Hunter Group operated in the city and the province of Parma. Later it was transferred to the Veneto and, finally, to southern Piedmont, where it remained from July 1944 until the end of hostilities.[4]

The RAP was established in July 1944. The men that formed these units were known as 'Arditi' and 'counterinsurgents' with specific training for waging the anti-partisan struggle. RAP troops operated both inside and outside cities, and carried out extensive counterinsurgency sweeps as well as operations of a surgical nature. These included intelligence and *controbanda* activities, espionage (sometimes using false identities), and indirect counterinsurgency activities (law and order, military police, monitoring, road blocks, etc.). The unit operated in Piedmont continuously until the end of hostilities. In early May 1945, after an armed retreat into Ivrea province, it surrendered to American troops. The RAP's structure was that of a light brigade which had its origins in the Italian 'pacification' experiences in the Balkans and East Africa.[5]

Within the RAP were two elite units, the Arditi Officers Units (RAU), which proved themselves particularly effective at partisan repression. These units were formed exclusively of junior and senior officers and distinguished themselves by carrying out surgical and *controbanda* operations. They became expert in fighting the Resistance, and were especially adept at employing the insurgents ' own techniques, such as ambush, against them. These special units

3 Archivio Ufficio Storico Stato Maggiore Esercito (AUSSME), RSI, I-1, B. 41, «Diario Storico Centro Addestramento Reparti Speciali».

4 Ibid., B. 11.

5 Ibid., B. 6, «Grafico della ripartizione dei reparti autonomi dell'esercito repubblicano, dislocati in Italia, secondo la dipendenza, reparti speciali», 22 February 1945.

carried out counterinsurgency operations, both in urban and wilderness environments, in the Piedmont provinces of Asti, Cuneo, Turin and Vercelli.

Interrogations and Torture

Torture involves the use of physical and psychological coercion for the purpose of punishing, extracting information or confessions, or weakening prisoners' morale. However, in addition to the instrumental motivations for torture cited above, sources and testimonies collected in the course of later criminal trials reveal a subtle form of sadism among some of the participants, indicating that they experienced pleasure in carrying out their violent tasks.[6]

The Italian Social Republic did not shrink from employing torture, considering it necessary practice for squeezing information out of ideologically committed irregular forces such as partisans and anti-Fascists. Torture was not only the prerogative of police forces or politicized units, but was widely used by the army, the Apennine Hunters and the Anti-Partisan Group. Arriving at precise numbers of people who were tortured by anti-partisan forces is practically impossible. In light of this, the analysis that follows will be limited to describing the methodologies employed by the interrogators.

Before discussing methodology, however, it is necessary to establish a spatial framework: where did torture occur? From analysis of the available sources, it emerges that torture was often committed inside civil habitations or within the barracks of the anti-partisan forces. However, regardless of where the torture occurred the common denominator in all instances was the employment of extreme violence. Generally, interrogations and torture that took place within civil habitations occurred following routine document checks, patrols, or raids on places where a partisan presence had been reported. The activities inside civilian dwellings differed from those in barracks by the types of persons involved. In a home, even innocent civilians whose only fault was having ties of blood or friendship with the wanted person, could be victims of Fascist violence.[7] If weapons or subversive materials were found during the search or the suspect was captured, the soldiers could choose, at their discretion, whether to carry out the interrogation at the suspect's home or transfer the suspect to the barracks.[8]

6 Istituto Storico della Resistenza di Torino (I.STO.RE.TO.), C.S.A., N. 527, «Sentenza C.s.a. di Torino 10 luglio 1947», 64.

7 Ibid., 19.

8 Ibid., 59.

The instruments of violence were usually the same regardless of the location. Threats were followed by severe beatings that could cause the suspect to lose consciousness. The beatings could also be accompanied by the use of 'instrumentation' designed to make the pain more acute and cause permanent disfigurement.[9] When the interrogations were over, the victims were often paraded through the streets of the town for intimidation purposes before being led before the firing squad. It was a clear demonstration of force against the population, with the aim of making everyone understand what could happen to those who opposed the government of the ISR.[10]

When the decision was made to transport the prisoner to the barracks, the interrogation and torture proceeded at a slower, more controlled pace. First, the detention period was increased; the arrested person could be locked up for days or entire weeks. It was therefore necessary for the Army to prepare areas designed for the long-term incarceration of prisoners and interrogation. Second, the duration of the interrogation and the torture were longer. The interrogator could take all the time he deemed necessary, from a few minutes to several hours.[11]

The conditions in which prisoners were held varied from barracks to barracks. In some partisan diaries, the prisoners describe facilities that were liveable, where food was regularly served, and where it was possible to sing and receive books from outside:

> The following morning, I got up feeling cheerful and in the afternoon I was called by Col. Farina for the interrogation. [...] So after two hours of interrogation I returned to my friends in the prison, who asked if they had beaten me. I told them they had threatened to but hadn't hit me. So dinner arrived, I ate, then together with my friends we got to singing Partisan songs and some Alpine tunes until we went to sleep. The next day I got a visit from a young lady who had brought me some books to read and I was right away wrapped up in reading a book titled 'Come divenni esploratore' [How I Became an Explorer].[12]

---

9 Ibid.

10 I.STO.RE.TO., C.S.A., N. 514, «Sentenza C.a.s. di Torino 21 luglio 1947», 3.

11 The sources used show that in some cases the interrogations of suspects or partisans lasted as long as four hours, without interruption.

12 Sergio Elia, *La mia prigionia. Diario* (Cuneo: Vincenzo Iebole Editore, 1945). On 6 March 1945, the partisan Elia was executed as a reprisal. The last lines of his diary are addressed to his family. 'My dear parents, in just a few hours my situation has changed, I am about to reach our heavenly home. I have prayed to our Lord that he will take me to Him. Have courage and forgive me for all I've done. The Lord has already forgiven me, don't weep.' Ibid., 16.

Other holding facilities were dramatically worse. In Ceva, for example, the prisoners' situation was extremely uncomfortable: they were crammed into small quarters without sanitary facilities and teeming with fleas.[13] Even fouler-smelling premises, described below, were reserved for those who underwent torture, so as to constitute a kind of 'quarantine' from those who had not yet been interrogated:

> Towards noon we saw two more young men arrive. One of them was in bad shape, his face was all bloodied by the punches he'd received during the interrogation. They put him in a separate room from us, damp, cold, without a cot, without blankets, without food. The other had head wounds and a cut below his ear, and his overcoat was covered with blood.[14]

As more time was available in the barracks, interrogations could take place with greater 'calm' and the jailer could choose what technique to use to extract information. Typically, both the soldiers of the Alpine Hunters and those of the Anti-Partisan Group first used psychological torture, such as threats; only later did they resort to physical torture. Generally, threats were not only verbal and were often followed by acts of intimidation such as holding weapons to the head or throat. Threats continued once the prisoners returned to their holding room – soldiers would burst in, railing at the prisoners and shouting that they would soon be executed.[15]

If psychological torture proved to be insufficient, the counter-insurgents moved on to physical torture. Once again, the basic method of torture that was employed was beatings; one beating suffered by a particular prisoner is described as follows:

> During the night, at intervals, he was interrogated and tortured, and since he continued to refuse to answer the questions that were put to him, Buscarini, who boasted of being a former boxer, gave him such a powerful punch that he knocked him down, unconscious.[16]

Kicks and punches could be followed by more 'refined' tortures that included the use of particular means of coercion like ox whips, pliers, screwdrivers,

---

13 Ferruccio Iebole, *Partigiani, Martiri Liguri, Piemontesi e «Cacciatori degli Appennini»* (Mondovì: Edizioni AeC, 2005), 84.

14 Ibid.

15 I.STO.RE.TO., C.S.A., «Sentenza C.a.s. di Torino 10 luglio 1947».

16 Ibid., 63.

pieces of wood, lashes or bamboo canes.[17] With this treatment, the soldiers generally succeeded in obtaining information. But, if the counter-insurgents became convinced during the interrogations that the prisoner had given only a partial confession, they would continue to inflict further pain with greater vigour and violence to complete the information gathering.[18]

Women who supported the Resistance movement, whether openly or in secret, were not spared from these practices. One emblematic case is that of Margherita Mo 'Meghi', a courier and coordinator of the Second Langhe Division. On December 7, 1944, she was stopped at one of the Apennine Hunters' checkpoints at San Michele di Mondovi and threatened with immediate execution by Lieutenant Rizzo. However, it was decided to interrogate her and to spare her life so as to be able to use her in gathering information against the partisans (which Mo never provided).[19] Immediately following her arrest, Mo was

---

17 Ibid.

18 The trial documents show that often it was also troops who abused the prisoners. The testimonies collected by the Extraordinary Courts of Assizes show the terror for the return of the troops from their time off duty – often drunk – who would enter the detention facilities and abuse the prisoners both verbally and physically. 'On the other hand, however, every evening the troops returned to the barracks (often drunk), and I suffered every sort of outrage: knives and pistols held to my throat, death threats, obligatory boxing matches. They delivered the blows and we were the ones on the receiving end, of course.' Iebole, *Partigiani, Martiri Liguri, Piemontesi e «Cacciatori degli Appennini»*, 85.

19 A similar attempt was made in relation to a female Partisan with the assumed name 'Tuska'. I became aware of the episode in the course of research done at the historical archive of the Resistance in Cuneo. Detained by the Apennine Hunters, the woman was interrogated for several days to get information on the Partisan bands in the area. The officer in charge of the interrogation was Col. Farina, who on several occasions tried to convince the woman to confess and join the ranks of the Fascists:

'– Your cousin told us that you collaborate with her to get information. Is that true?
– Actually what I did was very little, and I try to be as humble as possible, but I'll try to do more.
– I don't know that what you say is true, says Lieutenant Farina. I don't know if you are truly with us. Anyway, prove it. You can begin immediately: there are Partisans here in prison with you who are convinced and others who are less convinced. I want to know who the most hardened ones are, where their relatives who aren't at home are, certainly among the bandits ... Get them to talk and report to me everything you find out.'

Now I'm in a real fix! I could never accuse anyone, much less a Partisan ... How am I going to get out of this dilemma? ... Will I be clever enough? But I manage to reply ...:
– I'll try!
– Living in the Langhe, you can find out I lot of things that interest us.
– I have to distinguish. I'm not a partisan, I've never been together with them, so I don't have inside information ... but I'll see ... I know where they are, how they live.
– Do you know names?

first taken to the nursery school where the command was headquartered and then later to prison. Here the interrogation took place with the same methods of threats and violence used against men. Mo continued to deny participating in the resistance activity until Lieutenant Rizzo showed her a photograph of herself alongside a partisan, pistol in hand. At that point, the firing squad seemed inevitable and Lieutenant Rizzo had her head shaved, telling her. "This way you'll die uglier!" A few minutes later, however, an order arrived to stop the execution, thereby saving the partisan's life.[20]

Conclusion

Throughout the course of the civil war, the practice of torture, although never officially justified, was an instrument habitually used by the military and police units of the Italian Social Republic. It was a form of violence used to achieve two precise goals: eliminating the partisan movement, and completing the process of moral and physical elimination of the adversary through his complete dehumanization. The body of the prisoner became the test bench upon which to experiment with the prisoner's threshold of suffering. The body and mind of the adversary were totally at the mercy of the perpetrators. The torture of the ISR was a form of violence that, at the physical and psychological level, can find a first general explanation in Fascist awareness of failure to succeed in eliminating the Partisan movement. Thus, if the partisan bands could not be destroyed, it was preferred to attempt to strike, and destroy, the single individual.

Archival Sources

AUSSME (Historical Archive of the Army General Staff – *Archivio Ufficio Storico Stato Maggiore Esercito*). Catalogues: RSI, I-1, Folders (*Buste*) 6. 11, 41.

I.STO.RE.TO. (Historical Archive of the Italian Resistence Movement in Turin – Istitito Storico della Resistenza di Torino). Catalogues: C.S.A., N. 514, 527.

– Names, yes, lots of them, but they're noms di guerre … I don't know any of their real names.
– There's a radio transmitting station. Do you know where it is? They broadcast so badly that they should be ashamed … ', Tuska, 'Tuska. Maestria partigiana', *"Sempre in gamba né". Storie e personaggi della VI divisione Cion Silvio Bonfante e la II Zona Liguria*, ed. Ferruccio Iebole (Mondovì: Edizioni AeC, 2010), 118–119.

20 Ibid., 245.

Bibliography

Baldissara, Luca and Paolo Pezzino, eds. *Crimini e memorie di guerra*. Naples: L'ancora del mediterraneo, 2004.

Chiarini, Roberto. *L'ultimo fascismo: Storia e memoria della Repubblica di Salò*. Venice: Marsilio, 2009.

Del Boca, Angelo. *Italiani, brava gente?*. Vicenza: Neri Pozza Editore, 2005.

Elia, Sergio. *La mia prigionia: Diario*. Cuneo: Vincenzo Iebole Editore, 1945.

Fioravanzo, Monica. *Mussolini e Hitler: La repubblica sociale sotto il Terzo Reich*. Rome, Donzelli Editore, 2009.

Fulvetti, Gianluca and Francesca Pelini. *La politica del massacro.* Naples: L'Ancora del mediterraneo, 2006.

Gagliani, Dianella. *Brigate nere: Mussolini e la militarizzazione del Partito fascista repubblicano.* Turin: Bollati Boringhieri, 1999.

Iebole, Ferruccio. *Partigiani, Martiri Liguri, Piemontesi e «Cacciatori degli Appennini»*, Mondovì: Edizioni AeC, 2005.

Oliva, Gianni. *I vinti e i liberati: 8 settembre 1943-25 aprile 1945.* Milan: Mondadori, 1994.

Pisanò, Giorgio. *Gli ultimi in Grigioverde.* Milan: FPM, 1982.

Ranzato, Gabriele, ed. *Guerre fratricide. Le guerre civili in età contemporanea* Turin: Bollati Boringhieri, 1994.

Rocco, Giuseppe. *Con l'onore per l'onore: L'organizzazione militare della RSI sul finire della Seconda Guerra Mondiale.* Milan: Greco & Greco, 1998.

Rovatti, Toni. *Leoni Vegetariani: La violenza fascista durante la RSI.* Bologna: Clueb, 2011.

Tuska. 'Tuska. Maestria partigiana'. "Sempre in gamba né". *Storie e personaggi della VI divisione Cion Silvio Bonfante e la II Zona Liguria*, edited by Iebole Ferruccio, 118–119. Mondovì: Edizioni AeC, 2010.

CHAPTER 9

The Goli Otok Camp: Torture Justified by External Threats?

Martin Previšić

Abstract

After the Tito-Stalin split in 1948, Yugoslavia was faced with many challenges, not the least of which was to figure out how to survive without the abundant Soviet help upon which Yugoslavia had been relying since 1945. Members of the Communist Party who supported Stalin (and the Informbiro Resolution) posed a major problem as the establishment feared they might turn into enemies and help the Soviets in overthrowing Tito's regime should the Soviets take military action and invade Yugoslavia. In order to prevent a potential takeover from the inside, the KPJ (Communist Party of Yugoslavia) set up a system of concentration camps between 1949 and 1956 in which some 16,000 Stalinists were incarcerated. Most of them, about 13,000, were incarcerated in the biggest and most notorious top-secret prison and labour camp on Goli Otok*. It operated through a special system based on torture as a key method of imposing duress aimed at forcing the Informbiro members to denounce the members who were still free. Detained under false accusations, they were tortured to confess that they were Stalin's supporters and promise that they would restore their loyalty to the Party. Later on, Goli Otok generated heated debates over the justifiability of the violence committed.

Keywords

Goli Otok Camp (Barren Island) – Yugoslavia – Communist Party of Yugoslavia – torture – Cominform – UDBA – Josip Broz Tito

The break-up between Josip Broz Tito's Yugoslavia and the Soviet Union in the summer of 1948 was a result of Yugoslavia's independent foreign policy in the Balkans, which Tito did not see as detachment from Soviet foreign policy or the

---

* Since the time of completion of this article, my research grew into a book on Goli Otok labour camp, published in February 2019. (Martin Previsic, *Povijest Golog otoka*, Zagreb: Fraktura, 2019)

 | DOI:10.1163/9789004408012_011

general goals of the Communist movement, worldwide. The Tito-Stalin split was by no means a coincidence. Rather, its roots can be found in the widely different views on the post-war relations in Europe and the world. Yugoslav communists looked at the future of the Balkans with optimism and confidence. Enriched by their experiences from the four-year war they had waged at a great sacrifice, they hoped that they would contribute to the communization of the entire Balkan Peninsula and possibly a large part of Europe. The cause of the Tito-Stalin split was a result of Yugoslavia's interference with the civil war in Greece, its initiatives towards the unification with Bulgaria, and its actions in Albania. Unable to control Tito and the Yugoslavs, Stalin used various methods in an attempt to straighten out Yugoslavia and bring it back into the Cominform fold. Finally, by way of Cominform (the umbrella organization of the Communist parties) and its Resolution of 28 June 1948, Stalin openly attacked Yugoslavia, expelled it from the Cominform and called on the KPJ (Communist Party of Yugoslavia) to carry out a coup and overthrow Tito. Before long, the KPJ was divided by factional fighting between those who supported Tito and others in favour of Stalin, the latter called *ibeovci* (Informbiro supporters). In fear of possible military intervention as well as economic and diplomatic isolation, together with an upheaval at home, the Yugoslav leadership decided to isolate both the actual and potential IB supporters in prison and labour camps, the largest and most notorious being Goli Otok (literally, Barren Island).

The Goli Otok labour camp was set up in July 1949 by the federal UDBA (Yugoslav Secret Police), with the logistical assistance of its Croatian branch, as a response to the ever-growing diplomatic, economic, political, and military pressure from the USSR and IB countries after the IB Resolution proclaimed in June 1948.[1] The camp was just one segment of the extensive plan for interment of IB supporters, but it was also the largest of all detention centres, as 75% of the arrested and punished IB supporters ended up being interned there.[2] A total of 15,737 persons were incarcerated by the system, of which 13,000 were on Goli Otok.[3] Other camps and prisons were located in Bileće, Ramski Rit, Ugljan, Zabela (Požarevac), etc. In the camp, convicts were incarcerated in different compounds, depending on the mode of punishment and gender. The first group included administratively convicted persons (civilians). The second consisted of persons tried at regular or military courts. The third was intended for women.

1 Ivo Banac, *Sa Staljinom protiv Tita* (*Zagreb: Globus 1990*), 229–239; Martin Previšić, Broj kažnjenika na Golom otoku i drugim logorima za ibeovce u vrijeme sukoba sa SSSR-om (1948.-1956.) *Historijski zbornik* LXVI (2013): 177.

2 Previšić, Broj kažnjenika, 178–180.

3 Ibid.

The main purpose of the Goli Otok camp (1949–1956) was to "politically reform" the interned supporters of the Informbiro Resolution. This policy stemmed from the belief of the KPJ leadership that they should treat their disloyal members with more tolerance and forgiveness ('just bang them in the head, not make them shorter by the head') than their other opponents who did not share the same ideology.[4] The latter included collaborators and other "enemies of the state" who were dealt with in the cruel Stalinist clashes after 1945. The methods used at Goli Otok, however, did not deviate much from the Soviet model from which Yugoslav communists were trying to distance themselves. From the moment of their arrest, IB supporters were doomed to end up in prison, most probably at Goli Otok. During their interrogations in prisons all over the country, the prisoners in remand were forced to confess their IB activities and denounce other potential IB members who were still free.

People were violated long before they came to the camp. It all began in the remand prisons. UDBA investigators used violence, the degree of which depended on the prison in question, but also on the attitude of both the investigator and the detainee. The investigation itself lasted as long as the investigators deemed necessary to achieve their goals – on average three months, but sometimes as long as a year. The conditions in prisons were poor: cells were overcrowded, food was both bad and scarce, and hygiene minimal, which doubtlessly contributed to the anxiety felt by the arrested IB supporters.[5] After the investigation, they stood provisory, administratively conducted trials where various Party or UDBA committees usually sentenced them, most often to 24 months of community service.[6] Around 70% of IB supporters were punished in this way and then they were interned in camps or prisons.[7] The rest (30%) stood real trials and were sentenced to long prison terms. These were mainly commissioned officers of the Yugoslav Army who declared themselves in favour of the Informbiro Resolution. A total of 10,999 persons received administrative sentences and 4,340 were sentenced by courts.[8] With the exception of the post-war trials

4 Martin Previšić, History of the Goli Otok Cominformist Prison Camp 1949.-1956. (PhD diss., University of Zagreb, 2014); Milovan Đilas, *Vlast i pobuna*, Zagreb: EPH-Novi liber 2009, 272.

5 Đuro Bilić, *Goli otok i Dabravine* (Zagreb: 1998), 61–116.; Mihailo Simić, *Na Golom otoku* (Beograd: BIGZ 1990), 11–37; Dragoslav Mihailović, *Goli otok* (Beograd: Politika, 1990), 302–315.

6 Martin Previšić, Suđenja i kažnjavanja pristalica Rezolucije Informbiroa, *Hereticus* XI (2013): 197–214; Berislav Jandrić, *Hrvatska pod crvenom zvijezdom* (Zagreb: Srednja Europa, 2005), 296–301.

7 Ibid.

8 Hrvatski državni arhiv (HDA – Croatian State Arhive), RSUP SRH SDS INFORMBIRO, fond 1561, „Popis".

of collaborators and the so-called "public enemies", the authorities avoided big public trials.[9]

It needs to be pointed out that life on Goli Otok was imbued with violence. The trip alone to Goli Otok gave the IB supporters a sense of what treatment they could expect. The hours-long, sometimes even days-long trip, was spent in jam-packed, sealed carriages in which the prisoners had to relieve themselves. Upon arrival at the port of Bakar, they were transported by boat to Goli Otok, where the real violence started. While aboard the boat, the IB prisoners were thrown into the cargo-hold, which ended in broken arms and legs among many of the prisoners. The journey to the island took most of the night. In addition to the violence and uncertainty, the prisoners were terrified that the boat would sink. Inmate Ivan Kolundžić described the atmosphere during the transport to Goli Otok:

> We were stuffed up in cattle wagons and the train took us to Bakar. The railway station in Bakar was located on the seafront, where the "Punat" cargo carrier was already waiting. The total of 980 prisoners from Croatia, and I can say that all of them were the most loyal sons of our Revolution, embarked on that boat and we set off to Goli Otok right away. While aboard, many of us thought that they would drown us and that the press would report it as wreckage. But we eventually reached the island and that is where Tito showed us the true colours of his humanity…[10]

Unfortunately, upon arrival at Goli Otok, the prisoners were in for some new and unpleasant surprises. Running the gauntlet of inmates upon arrival was the most memorable moment for the majority of IB supporters. They were forced to walk between the lined-up inmates who were beating them. After this sadistic ritual, most prisoners were badly injured, while some even succumbed to the injuries thus inflicted. The testimony of prisoner Vlado Bobinac evokes the atmosphere:

> I was pushed into the gauntlet, which was very long, all across the camp, with 2,500–3,000 people in it. They were lined up in two columns facing each other and we were supposed to walk in between. The gauntlet stretched from the port, across the wired area of the main square and all

9 Boro Krivokapić, *Dahauški procesi*, Beograd 1986; Igor Torkar (Boris Fakin), *Umiranje na rate*, Zagreb 1984; Dragan Marković, *Istina o Golom otoku*, 139–147; Slavko Ćuruvija, *Ibeovac – Ja Vlado Dapčević*, (Beograd: Filip Višnjić 1990), 147.

10 Unpublished memoir Ivan Kolundžić.

> the way up to the last barracks. All I heard were roars: "Whack the bandits, whack the bandits!" Imagine the crowd of 2,500-3,000 just roaring like that. While running through the gauntlet, I was brutally beaten as they savagely beat us with whatever they had. The whole gauntlet path turned red from our blood. When I reached the end of it, I was covered with blood, hardly conscious. We arrived at the end of the wall, a lot of us, and we were then forced to stand there with heads bowed and arms behind the back, while blood was gushing from our noses, heads, arms and legs. When the last newcomer ran through the gauntlet, the roaring stopped. Only here and there you could hear someone shout "Whack the bandits" and he would kick you at that.[11]

Soon, the prisoners had to get used to the camp regime and the demands that UDBA expected them to obey. First of all, they were supposed to realize that it was not a regular prison where your freedom depended on the prison term to which you were sentenced. Rather, it was a place of torture where meeting UDBA's criteria in the *process of political re-education* was the only way to resocialize and be free again. Specifically, it meant that all convicts were expected to confess publically that that they had supported Stalin (so-called *revision of views*). Furthermore, they were expected to denounce alleged IB supporters who were still free, but also to identify those in the camp. If there were none, they had to make them up. Participation in the violence against other prisoners was rewarded and viewed by the camp authorities as the politically correct attitude towards the 'bandits', as those who still did not revise their opinion were called. In simple terms, the inmates were forced to collaborate in different acts of violence. These included urinating on prisoners; senselessly carrying rocks over from one pile to another; collective beating of a chosen prisoner on a daily basis; deprivation of sleep, food and cigarette rations. These acts were the core of the *political re-education process*. The rise or fall of prisoners in the camp hierarchy depended on how and to what degree they were meeting the imposed conditions. Alfred Pal, a former prisoner, who spent four years at Goli Otok, best describes it:

> At Goli Otok you had to declare that you were against Informbiro and Stalin, which most of us did. That was the first step. Then one of UDBA's men would say: 'Good. It's only words. Now prove that you are against Stalin by finding an enemy here at Goli Otok. Go and talk to other prisoners and

---

11 Testimony given to the author: Zagreb – May 2010, October 2011.

> report on those who are in favour of Informbiro.' Naturally, most of the prisoners would then strike up a conversation with others, lure a person into saying something that he shouldn't have said and then he would be reported to the investigator. The UDBA man would say: 'OK, but it's still only words. You must prove with your acts that you are against Stalin. Go and beat some Stalin supporters tonight.' If the prisoner refused to do so, he himself would be beaten. The UDBA man then said: 'So, that's it! You are all for our Party and yet you wouldn't beat our enemy!' And the beating would continue. That was the system which was breaking people down.[12]

The Goli Otok camp was organized along the line of pseudo self-governance, where inmates controlled other inmates, without direct physical interference on the part of UDBA, but still under UDBA's control. The establishment of the fictional self-government system commenced when a group of prisoners from Sarajevo Central Prison called "Bosnians", prepared in advance by UDBA, arrived in late July of 1949.[13]

In many ways it was a copy of the systems adopted in the Nazi camps or in Gulag. In reality, neither the UDBA nor the guards were physically present within the "Wire" i.e. the fenced-off area where the inmates lived. Neither were they personally involved in physical violence. The guards who observed the camp from their machine-gun nests around the "Wire" were the official staff closest to the prisoners. Moreover, there was a pyramidal system where prisoners were classified in different groups in accordance with the extent and quality of their cooperation. Their position in this hierarchy determined their quality of life. The highest echelon of the camp self-governance was the *Center.* It was directly answerable to the UDBA and was tasked with implementing their directives. The "Center" was headed by the "chief". He was the most important person from the ranks of prisoners in the camp and in the self-government system, a person of UDBA's particular trust. In some cases this position was held by undercover UDBA members and they had unlimited powers. The second level of the camp self-government system was the barracks leadership or pavilion staffs, one in each barrack/pavilion. Pavilion staff took care of the day-to-day functioning of their respective pavilions: distribution of food, drinks and cigarettes, haircuts, laundry, formation of labor groups, etc.

The inmates were also divided into categories: for instance, *activists* were those who revised their views (changed their political opinion). They had the

12 Testimony given to the author: Zagreb-September 2009.

13 Dragoslav Mihailović, *Goli otok 1*, 208–227.

highest status in the camp and were candidates for the leading positions in the camp's self-governance. The *boycotted* ranked the lowest, as they did not revise their views.

Boycott was the most effective means of pressure against prisoners to make them change their political views. The boycotted prisoners wore different clothes so that they were physically distinct from other inmates. Also, it meant that other inmates were free to beat them. The boycotted wore red stripes, after which they were nicknamed "the generals"; black shirts, which symbolized Fascism; or they wore T-shirts inside out. They were not allowed to talk to anyone, and always had to keep their heads down. In the camp, they were also called "bandits", although outside it was a common name for all imprisoned IB members. Very often they were not allowed to stand in line on the arrival of new convicts. They were not entitled to daily rations of cigarettes or extra rations of food. During half of the allowed time for sleep, i.e. four hours, the boycotted had to *watch the bucket*. The buckets were placed outside each pavilion to be used as toilets, but it was forbidden to use them during the night. The boycotted were assigned to watch the buckets full of faeces for four hours with their heads down and faces over the buckets.

The activists exposed them to the most severe pressure and violence. In this way, they were proving their loyalty to the Party. Also, activists were assigned easier jobs – those of cooks, gardeners, water carriers, food servers – or held some functions in the self-governance bodies, such as cell leaders, hygienists, cultural or educational officers, etc. The *boycotted*, on the other hand, were forced into hard labour, most often in the quarry, where they were regularly beaten.

The Goli Otok camp was not set up on the model of other camps, but it bore close resemblance to the Nazi or Soviet labour camps. Its creation was mostly based on the accounts of the Yugoslavs who had formerly been incarcerated in the USSR or in Nazi camps during the Second World War. A comparative analysis indicates that Goli Otok had a similar organizational structure as the Nazi camps: the real camp administration (SS – UDBA) and the pseudo administration composed of camp inmates who were directly subordinated to the real administration (Lagerälteste, Blokälteste, Kapo – Chief of Centre, cell leaders, work supervisors).[14] The system of privileges copied, to a great extent, the *Gulag* in the USSR where the privileged inmates were given easier jobs, whereas those with no privileges had to do all the hard labour (felling of trees in Gulag – labour in the quarry at Goli Otok).[15] Likewise, it seems that the system of

14 Previšić, History of the Goli Otok.

15 Ibid.

denunciating the people who were still free in order to earn some privileges (Gulag) strongly influenced the system of the political re-education process at Goli Otok.

Goli Otok was not just a place where senseless activities were carried out. The labour had a twofold purpose: to turn a profit and political re-education. Already in the first year of its existence, the economic life on the island was organized in a lucrative and productive way, with a whole range of production plants unified in the *Kombinat Mermer* (Marble Mill), managed by the UDBA administration. It included wood, stone, gravel and metal processing plants. The final products ranged from various wooden, stone and metal items, such as cigarette cases, mouthpieces, furniture, plates, concrete flowerpots, monuments, metal constructions. Boats were repaired in the small shipyard, and the camp itself had a small fleet of boats used to transport people, materials and goods. The fleet was also maintained for security purposes. Here too, violence was used to demonstrate the status of inmates in the internal hierarchy: those who would not collaborate with UDBA were assigned harder jobs. The most stubborn supporters were forced to cooperate with UDBA by forcing them to perform senseless tasks. For example, in the summer, prisoners had to carry rocks from one pile to another under the hot Adriatic sun or dig in the quarries; in the winter, they were set to work extracting sand from the cold sea. The idea was to break the prisoners down by putting them to hard labour so as to make them revise their beliefs. This violence would stop when the prisoners did what UDBA expected them to do. They would then get much easier assignments, e.g. making cigarette cases, chairs and tables, or repairing clothes.

The convicts' labour changed the appearance of both the camp and the island. Intentionally or not, the camp administration and, consequently, the prisoners transformed that barren, treeless island and partly reforested it. In addition, they erected a small semi-industrial complex. Various punishments imposed on prisoners as a means of pressure against the IB members ("revision of views") had tremendous effects on the appearance of Goli Otok and its eco-system.[16] For example, one of the methods of punishment was to force inmates to protect seedlings with their own bodies (it was just one of the methods of reforestation). This resulted in 60 hectares of the island's total of 470 becoming reforested by the time of the prison's closure.[17]

---

16 Previšić, Martin, Prokić, Milica. Ekohistorijski aspekti proučavanja logora na Golom otoku 1949.-1956. Ekonomska i ekohistorija, Časopis za gospodarsku povijest i povijest okoliša. Vol XII.

17 Ibid.

Poor hygiene and various diseases were typical for all camps and prisons, and Goli Otok was no exception. Until 1952, there were no sanitary facilities on Goli Otok. The inmates did not bathe, except occasionally in the sea. As the barracks were not cleaned, they were full of parasites, such as bedbugs and lice. Inmates never changed clothes and were reduced to relieving themselves in a common pit. The constant stress, physical abuse and poor nutrition caused several diseases, most often inflammations, gastrointestinal infections and concussions. Particularly dangerous was the outbreak of typhoid in 1951, which took 150 lives. The outbreak did, however, result in some improvements in living conditions in the camp.[18] Prisoners were medically treated by inmate-physicians who worked under the supervision of UDBA's official physicians. Nevertheless, despite medical attention, prisoners died due to a variety of causes – violence, disease, suicide. The dead were first buried at the provisory graveyards on the island and later the bodies were transported mainly to the cemeteries in Zagreb or Rijeka.[19] A total of 400 interned IB supporters died in all camps and prisons, half of them killed by typhoid in 1951.[20] Inmate Vlado Bobinac recalls that the outbreak of typhus took a large number of lives:

> The epidemic of spotted fever broke out in 1951. We were fenced in wire and did not go to work for about ten days. We would take off the rags we wore and kill louses with the pieces of paper from the bags in which cement was kept. During that period we were always outside, we only slept in the barracks. Dr. Nikolić did a great job in combatting that epidemic. Among other things, he ordered that an old caldron be turned into a steamer where clothes were thrown so as to kill the pest. Some people died. There was also an epidemic of infectious jaundice and tuberculosis.[21]

From 1950 until 1954, there was a strictly isolated department of the Goli Otok labour camp: Worksite 101, informally called *Petrova Rupa* (Peter's Pit).[22] This special section of the camp was notorious for the particularly brutal treatment

18 HDA, RSUP SRH SDS INFORMBIRO, fond 1561, Umrli – brojčani podaci po godinama 1949–1986.

19 Dragoslav Mihailović, *Kratka istorija satiranja*, (Beograd: Zavod za udžbenike 2011). 66.

20 Ibid.

21 HDA, RSUP SRH SDS, fond 1561, Dosje Milovan Zec.

22 Previšić, History of the Goli Otok; Dragan Marković, *Istina o Golom otoku* (Beograd; Partizanska knjiga, 1987), 113–119; Nikola Golubović, *Moje nemjerljivo blago* (Beograd, 2013), 71–116.

the guards meted out to the inmates. The barracks were built in the pit where bauxite ore had been dug before the Second World War. Some 130 of the most fervent IB followers, predominantly communists of Tito's generation, participants in the October Revolution, founders of the KPJ, emigrants in the USSR, losers in factional struggles in the KPJ, partisan leaders, and highly ranked officers of the Yugoslav People's Army (JNA), were all held in the barracks. Out of fear that they could come into power should Tito fall, UDBA kept them in isolation and in extremely harsh conditions marked by violence and hard labour.

Among the prisoners in Petrova Rupa, some were very influential prior to their arrest and imprisonment: Dragutin Gustinčić, one of the founders of the KPJ and member of the CK (Central Committee) KPJ; Nikola Kovačević, participant in the October Revolution and in the 1936–1939 Spanish Republican Army; Adolf Štumf, official in the Kominterna apparatus; Blažo Raičević, secretary of the Montenegrin Provincial Committee and Yugoslav ambassador to Poland after 1945; Mirko Marković, who spent nine years in the USSR and attended several political schools there, such as the KUNMZ (Communist University of the National Minorities of the West), a high ranking officer in the American battalion of the "George Washington" International Brigade in the Spanish Civil War and later professor of political economy after the war; Dimitrije Stanisavljević – Krka, who worked in the USSR as an NKVD official, fought in the Spanish Civil War as political comissar in the "Dimitrov" Battalion; General Veljko Žižić (Military Court Chief Prosecutor); generals Branko Poljanac, Branko Petričević and Đoko Mirašević (Belgrade Military District Chief of Staff); Dr. Savo Zlatić, who established the Partisan Hospital at Petrova Gora and headed the Federal Ministry of Industry after the war; Milan Kalafatić, member of four Communist parties (in Yugoslavia, Spain, France, and the USSR) and also lieutenant in the Spanish Republican Army; Vuko Tmušić and Niko Pavić, ministers in the government of Montenegro; Ludvik Krzel, Slovenian writer, a former Dachau prisoner; and Vicko Jelaska, Tito's factional opponent in the 1930s.

Once the prisoners satisfied all of the demands imposed on them in the process of political re-education, they were set free. Prior to discharge, most of them had to sign a document obliging them to keep collaborating with UDBA and to closely watch the IB supporters who were free. In addition, they committed themselves to the code of silence with regard to Goli Otok and the events that took place there. Inmate Milovan Zec was forced to sign such a document:

> I shall report to UDBA organs on subversive activities and carry out all tasks imposed upon me in the interest of the people. I shall not talk to anyone about this cooperation or give ground to anyone to find out about

> this cooperation. Should I renege on this commitment, I am prepared for the severest punishment by the law.[23]

Upon their discharge, the former prisoners kept experiencing all kinds of pressures, including public condemnation. No one wanted to be in the company of an ex-convict from Goli Otok. Besides being in disfavour with the administrative bodies, the prisoners had serious difficulties finding jobs or places to live, and in exercising other rights. In fact, the violence never stopped. The important criterion that determined the treatment of returnees was their further collaboration with UDBA in spying on other returnees, former inmates. Those who had not been good collaborators with UDBA at Goli Otok were more closely followed after their release and placed under stricter surveillance. By way of illustration, in 1951 the Croatian branch of UDBA had as many as 2,422 agents and informants tasked with following IB supporters – returnees from the prisons and camps.[24]

In addition, the system, i.e. the Party, controlled IB members as a group, by assessing the level of threat they posed to the system, but also assessing the possibilities of their inclusion into the political life of the country. Three years after the conflict between Yugoslavia and the USSR had ended, on 17 May 1958, the highest authorities of the Yugoslav Party reached a decision demanding that the federal UDBA conduct a collective analysis of the former IB members, including both persons detained in the camps and those suspected of being Stalin's supporters, and to classify them in one of the four categories depending on the level of threat they pose to the regime.[25] UDBA then came up with four categories according to which the analysis was conducted:
- Category 1: persons who toed the line of the SKJ, revised their former IB oriented views and therefore could be members of the Party;
- Category 2: persons who toed the line of the SKJ (the active ones), revised their former IB oriented views, but could not be members of the Party as yet because they did not actively participate in the work of the socio-political organizations;
- Category 3: (the passive ones) persons who revised their political views, but did not take part in the socio-political organizations;
- Category 4: persons who still held their IB views and had to be politically isolated.

23 HDA, RSUP SRH SDS, fond 1561. Dosje Milovan Zec, Obaveza.

24 Previšić, History of the Goli Otok.

25 HDA, RSUP SRH SDS INFORMBIRO, fond 1561, Elaborat Osvrt na problem rehabilitacije bivših pristalica Rezolucije Informbiroa.

In 1958, a total of 26,244 persons were included in the analysis based on these criteria:

- Category 1: 4,590 persons;
- Category 2: 6,521 persons;
- Category 3: 11,923 persons;
- Category 4: 3,210 persons.

The number of persons grouped in categories 2 and 3 indicate distrust towards IB members (or their indifference to politics), but also potential for gradual rehabilitation. Doubtlessly, the IB members were under considerable suspicion upon their return from the prison and during the conflict with the IB. However, already in 1958, there was a clear tendency (and that was part of the SKJ policy at the time) for rehabilitation of the former IB members, their political reactivation and inclusion in the Party. The completed *differentiation* resulted in the very possible return of 4,590 former IB members to the Party and another 6,521 persons who were on the verge of fulfilling the requirements for the inclusion in the SKJ. The policy towards the IB members sought to include as many of them as possible in the political life, albeit only at the level of membership. This can be seen from the analysis conducted five years later and the survey from 1963.[26]

- Category 1: 5,537 persons;
- Category 2: 9,809 persons;
- Category 3: 8,575 persons;
- Category 4: 1,903 persons.

The ratio between the numbers in the first two categories in 1958 and in 1963 shows an increasing tendency towards the rehabilitation of the former IB members, i.e. their fulfilling of the preconditions for inclusion in the Party. Of course, this was a demanding and slow process. In 1963, a total of 5,527 persons applied for membership in the Party, whereas 3,759 were actually admitted.[27] The Party exerted strict control over IB members, but in the late 1960s it predominantly focused on individual cases and extremist political groups. Nevertheless, IB members were looked upon with suspicion and caution until the very end of Tito's Yugoslavia. Stigmatized as camp prisoners and traitors for the rest of their lives, IB members were always deeply aware of this distrust.

To this day, the discussion of the Goli Otok labour camp is something of a taboo. In the collective memory, it has remained a place remembered for its brutal violence used against the inmates in the process of political re-education. There have been attempts to justify the existence of the camp as a place of

26 Ibid.

27 Ibid.

anti-Stalinist Stalinism, citing the need to defend the country from external threat.[28] The specific features that distinguish Goli Otok from other prisons were the methods of violence and pseudo self-governance used in the process of political re-education combined with the policy of rewarding prisoners who were compliant with the re-education process. In view of the above, the Goli Otok labour camp deserves to be thoroughly considered and analysed by historiography and addressed from different angles for what it can teach us about the nature of political oppression. In the same vein, it teaches us about what can happen when a system, regardless of its ideological predispositions, finds itself faced with the risk of collapse or a military intervention from the outside, a situation that Yugoslavia faced in 1949 when the camp was established. The history of Goli Otok also tells us about the fear of possible coup from the inside, which Stalin fervently hoped for after passing the Informbiro Resolution. Most of all, it tells us about the nature of Stalinism, both in Yugoslavia and in general. However, this camp should also be analyzed in the context of twentieth-century history, which was marked precisely by the existence of such camps. Surprisingly, given that the Goli Otok camp has a great significance in the history of the former Yugoslavia and Croatia, there is no memorial centre or museum of any kind on that island. The place that saw the torture of 13,000 people, and the deaths of 300 of them, has not been adequately marked. This is another indication of how society, in this case post-Yugoslav society, treats its own history.

Bibliography

Banac, Ivo. *Sa Staljinom protiv Tita*. Zagreb: Globus, 1990.

Banac, Ivo. *With Stalin against Tito – Cominformist Splits in Yugoslav Communism* (Ithaca-London: Cornell University Press 1988).

Bilić, Đuro. *Goli otok i Dabravine*. Zagreb: vlastita naklada [self-published], 1998.

Bobinac, Vladimir. Testimony given to the author: Zagreb, May 2010, October 2011.

Ćuruvija, Slavko. *Ibeovac – Ja Vlado Dapčević*, Beograd: Filip Višnjić 1990.

Dedijer, Vladimir. *Novi prilozi za biografiju Josipa Broza Tita, Tom 3*. Beograd: Rad 1984.

Đilas, Milovan. *Vlast i pobuna*. Zagreb 2009.

Golubović, Nikola. *Moje nemjerljivo blago*. Beograd: vlastita naklada [self-published], 2013.

Hrvatski državni arhiv Zagreb (Croatian State Archive, Zagreb Croatia).

28 *Praxis-Filozofski dvomjesečnik*. Br.3–4.1972, 381; Vladimir Dedijer, Novi prilozi za biografiju Josipa Broza Tita, Tom 3, 478.

Hrvatski državni arhiv Zagreb. RSUP SRH SDS, fond 1561. Dosje Milovan Zec, Obaveza.

Hrvatski državni arhiv Zagreb. RSUP SRH SDS INFORMBIRO, fond 1561, Elaborat Osvrt na problem rehabilitacije bivših pristalica Rezolucije Informbiroa.

Hrvatski državni arhiv Zagreb. RSUP SRH SDS INFORMBIRO, fond 1561, Umrli – brojčani podaci po godinama 1949–1986.

Hrvatski državni arhiv Zagreb. RSUP SRH SDS INFORMBIRO, fond 1561, „Popis".

Jandrić, Berislav. *Hrvatska pod crvenom zvijezdom*. Zagreb: Srednja Europa, 2005.

Kolundžić, Ivan. Unpublished memoir.

Krivokapić, Boro. *Dahauški procesi*. Beograd 1986.

Marković, Dragan. *Istina o Golom otoku*. Beograd: Partizanska knjiga,1987.

Mihailović, Dragoslav. *Goli otok 1*. Beograd: Politika, 1990.

Mihailović, Dragoslav. *Kratka istorija satiranja*. Beograd: Zavod za udžbenike 2011.

Pal, Alfred. Testimony given to the author: Zagreb, September 2009.

Praxis-Filozofski dvomjesečnik. Br. 3–4. Zagreb 1972.

Previšić, Martin. Broj kažnjenika na Golom otoku i drugim logorima za ibeovce u vrijeme sukoba sa SSSR-om (1948.-1956.), *Historijski zbornik* LXVI (2013): 173–193.

Previšić, Martin. Suđenja i kažnjavanja pristalica Rezolucije Informbiroa, *Hereticus* XI (2013): 197–214.

Previšić, Martin. History of the Goli Otok Cominformist Prison Camp 1949.-1956., PhD diss., University of Zagreb, 2014.

Previšić, Martin, Prokić, Milica. Ekohistorijski aspekti proučavanja logora na Golom otoku 1949.-1956. Ekonomska i ekohistorija, Časopis za gospodarsku povijest i povijest okoliša. Vol XII.

Simić, Mihailo. *Na Golom otoku*. Beograd: BIGZ 1990.

Torkar, Igor (Boris Fakin). *Umiranje na rate*. Zagreb 1984.

CHAPTER 10

'A Real Show of Horrors': Reading Representations of Torture in *A Clockwork Orange* and *American Psycho*

Stuart Molloy

Abstract

'A Real Show of Horrors' approaches torture through its representation in literature. The chapter compares Anthony Burgess' *A Clockwork Orange* (1962) and Bret Easton Ellis' *American Psycho* (1991) while juxtaposing the two novels' representations of torture with three canonical legal-philosophical assessments of the practice: Article 1 of the United Nations Convention Against Torture (1984); Michel Foucault's *Discipline and Punish* (1975); and Elaine Scarry's *The Body in Pain* (1985). After surveying the structure of torture according to these sources, discussion focused on the literature proceeds around the notion of instrumentality and the role of imagination. What does torture try to accomplish, and what are the implications of imagining it?

Keywords

torture – *A Clockwork Orange* – *American Psycho* – Article 1 – *Discipline and Punish* – *The Body in Pain* – instrumentality – imagination

A Clockwork Orange[1] and *American Psycho*[2] share their constitutions in common. Both are narrated in the first person by protagonists in many ways characteristic of the psychopath as the figure frequently appears in twentieth-century English fiction: white; male; highly intelligent; premeditative and remorseless; equally capable of charm and extreme violence. 'Alexander the Large'[3] – Alex – and Patrick Bateman, respectively, imagine and perform

1 Anthony Burgess, *A Clockwork Orange* (London: Penguin, 2000 [1962]).
2 Bret Easton Ellis, *American Psycho* (New York: Picador, 1991).
3 Burgess, *A Clockwork Orange*, 36.

 | DOI:10.1163/9789004408012_012

acts of organised brutality as a matter of routine, positing a 'real show of horrors'[4] as they explicitly report these acts to the reader. Most would agree that the characters' reports contain shocking if not sickening cruelty. But would there be a like degree of unanimity in categorising the content as torture? It is the complex issue of torture's definition which this chapter discusses, scope limited to the notion of instrumentality and the role of imagination. Must torture be goal-oriented, and what counts as a goal in torture? Must an act of torture be imagined before it can be performed? If so, what might the implications be? The discussion compares the two novels alongside three canonical legal-philosophical assessments of torture. The latter include: Article 1 of the United Nations Convention Against Torture;[5] Part One of Michel Foucault's *Discipline and Punish*, 'Torture';[6] and Chapter 1 of Elaine Scarry's *The Body in Pain*, 'The Structure of Torture: *The Conversion of Real Pain into the Fiction of Power*'.[7] To manage this substantial critical corpus, the key elements of each are surveyed in table form. Focus on the fiction, meanwhile, is limited to a single episode from each novel. The episode from *A Clockwork Orange* is the fourth chapter of Part Two. This features the narrator-protagonist subjected to the Government's experimental 'Reclamation Treatment'[8] – variously called 'Ludovico's Technique'[9] – in order to cure him of his criminality. The episode from *American Psycho* is the forty-second chapter, 'Girl', in which the narrator-protagonist variously brutalizes an anonymous young woman using a nail gun, a power drill, a starved rat, and a chainsaw. Finally, the conclusion draws on excerpts from mythology and modernism to suggest that torture, grasped through fictional representations, is bound to be unending ontologically despite the inherently interventionist aim of expanding on it epistemologically. However much torture is talked about legally and philosophically in a bid to shrink its practical presence in reality, no amount of theorisation will in fact excise it from human conduct.

---

4 Ibid., 76.

5 'Convention against Torture and Other Cruel, Inhuman or Degrading Treatment or Punishment 1984,' *United Nations*, accessed 25 February 2016, http://www.un.org/documents/ga/res/39/a39r046.htm.

6 Michel Foucault, *Discipline and Punish: The Birth of the Prison*, trans. Alan Sheridan (London: Penguin, 1991 [1975]), 1–70.

7 Elaine Scarry, *The Body in Pain: The Making and Unmaking of the World* (New York and Oxford: Oxford University Press, 1987 [1985]), 3–59 (inc. Introduction).

8 Burgess, *A Clockwork Orange*, 70.

9 Ibid, 62.

Following is the survey of the definitions of torture per the three canonical legal-philosophical sources:

TABLE 10.1 Three perspectives on the structure of torture

	UNCAT	Foucault[a]	Scarry[b]
	Torture is:	Torture 'must':[c]	Torture is:
Part 1	the intentional infliction of severe pain or suffering, physical or mental,	produce a certain degree of pain, measured exactly or at least calculated, compared and hierarchised	pain inflicted in ever-intensifying ways, its amplification corroborating its objectification,
Part 2	apparently motivated by the goal of obtaining information or confession; or exacting punishment; or disseminating intimidation,	in order to correlate corporal effect with criminal seriousness, i.e. torture is punishment,	which in turn denies it as pain
Part 3	by an authorised agent on behalf of the State.	and it must mark the body spectacularly – by amputation, evisceration, scarification, etc., i.e. torture intends control by intimidation.	to translate it as power, at least per the warped understanding of the torturer.

[a] Foucault in *Discipline and Punish* deals specifically with judicial torture in the West, common practice before and into the eighteenth century. UNCAT and Scarry, by contrast, deal with the kind of torture which came after judicial torture; that is, torture in its outlawed and fugitive form.

[b] The constituents of Scarry's view on torture are difficult to compartmentalise. Much of this has to do with Scarry's assertion that torture's constituents paradoxically occur both sequentially and simultaneously (19).

[c] Foucault, *Discipline and Punish*, 33. The imperative voice marks judicial torture's entitled and obligated nature.

Throughout these assessments is a common triplicate. 1) Each has a three-part structure; 2) the first part of that structure is the infliction or production of pain; and, 3) predominant is the view that torture is a demonstration of power by an authority, usually the substance of its third part. This last answers with uniformity the question of what counts as a goal in torture, and answers affirmatively that torture must be goal-oriented. Confession, punishment, and intimidation are also given mention, but they are not explicitly carried through each assessment uniformly. The literature agrees, but with modulation. The goal of torture is power and pleasure combined. That is, the pleasure of power. When a torturer is not an authorised agent of the State but an individual sadist acting purely on their own volition, as is the case with Alex and Patrick, denying that torture need be State-sanctioned, power is erotically charged. Within the libidinal economy of the sadistic subject power is bound up with pleasure, the one a means of obtaining the other. In accompaniment with the suffering of their victim, the sadist's pleasure is in dominating them, and carrying out torture provides an unequivocally dominant position.

That Patrick is a sadist has been strongly suggested already by the inventory of the implements with which the character dominates and destroys the young woman in 'Girl'. But that Alex is a sadist stands asserted without any evidence. Indeed, the decision to focus on the character's subjection to the Government's Reclamation Treatment obscures his sadism because it positions him as a victim and makes it hard to see him as victimizer. However, traces of his previous occupation of that empowered position still remain. As per the character of the Minister of the Interior, Alex is in prison as a 'common criminal', a 'vicious young hoodlum'.[10] Although Alex's commonness may be contentious, his viciousness is not. The character's two favourite pastimes prior to his arrest were listening to classical music and doing 'the ultra-violent'.[11] Limited in prison to fantasising about the latter, Alex takes to reading the Bible. This the character does not because he has arrived at a newfound concern for his ethical education and rehabilitation, as the prison chaplain assumes, but because he relishes the points of gruesomeness in what he calls 'the big book.'[12] About the Crucifixion Alex says:

> I read all about the scourging and the crowning with thorns and then the cross...While the stereo played bits of lovely Bach I closed my glazzies

10 Burgess, *A Clockwork Orange*, 69.

11 Ibid, 3.

12 Ibid, 60.

> and viddied myself helping in and even taking charge of the tolchocking and the nailing in, being dressed in a like toga that was the heighth of Roman fashion.[13]

Alex's delight in violence is vivid.

Mapped over torture's essential purpose as a demonstration of power are several ostensible motivations listed explicitly in Article 1: confession; punishment; and intimidation. Conveyed by their separation is the suggestion that these particularized explanations of goal-orientation are somehow discrete from torture's uniform aim. In fact none of them are so, but rather function within it. To extract confession, to exact punishment, to expound intimidation are all actions made possible by an empowered position, and indeed disclose exactly that. For Foucault, these three are profoundly supportive of one another, even mutually constitutive, combining in the formula: we punish you perhaps for nothing, but in subjecting you to punishment you'll admit to something; therefore it is indeed for something and not nothing for which we punish you; in this way we secure the State's legitimacy and your position outside its lawfulness, which did not begin with your condemnation, but was reason for us to condemn you. So, legitimately condemned, we punish you publicly with torture and at the same time make a threat out of you with which to keep control of the public: do what this person has done and this shall be done to you.

Current thinking counts such reasoning as solipsistic sophistry. But for those who articulated it, and for the majority of their contemporaries, operative in a previous episteme, it was perfectly acceptable penology. However surprising and objectionable such may seem now, in and before the eighteenth century there were 'few radical criticisms of torture'.[14] Rather, it was understood that '[a]trocity [was] that part of a crime that the punishment turn[ed] back as torture'.[15] Begun over the nineteenth century, the present episteme presupposes torture to be inhumane. Embedded in this 'modality of knowledge',[16] torture stands largely de-legalised and as such is viewed as both tool and sign of debased regimes. The rationale for torture, which remains unchanged, has therefore to be condemned precisely as sophistry by critics – as it is by Scarry, who deems confession, punishment, and intimidation 'false motives'[17] – while for

13 Ibid.

14 Foucault, *Discipline and Punish*, 40.

15 Ibid, 56.

16 Ibid, 28.

17 Scarry, *The Body in Pain*, 57–58.

those who utilize it – that is, torture and its rationale – the sophistry has to be ignored precisely through solipsism lest the façade of legitimacy collapse.

Accepted or condemned, proper or false, what has the literature to say about these drivers of torture? What little *American Psycho* says is tacit. Confession and punishment are ostensible indeed, surface aims which whether present or absent leave unchanged the core goal beneath: the demonstration of power, driven for the sadist by the pleasure of power's experience, of which intimidation is a part. Patrick is uninterested in forcing confession, and if the novel's protagonist is punishing the girl, there is no evidence in the text for what he is punishing her. Many read the character as a misogynist, a reading which does allow the logic of punishment to appear: Patrick tortures/punishes the girl simply because she is one of the hated female sex. But on balance a charge of misogyny is not an easy one to maintain. Patrick's narrative features thirteen incidents of violence given sustained description. The count of the narrator's victims over the course of these reaches twenty-three. One is a five-year-old boy; two are dogs, a mutt named Gizmo and a Sharpei named Richard; eight are women; and the remaining twelve are men, bringing the total number of male victims to thirteen, a clear majority.

A Clockwork Orange offers more. However illegitimate Alex's Government is presumed to be, because of its totalitarianism, it could legitimately claim its treatment of him to be punishment on account of his violent criminality. Yet it does not. Believing that '[p]unishment means nothing to them', the Government's position is that Alex and his ilk 'can best be dealt with on a purely curative basis.'[18] In the words of the Minister of the Interior again: ' "Kill the criminal reflex, that's all." ' In styling what is done to Alex as a cure, made plain from the outset by the word 'treatment', punishment is discredited as an obvious goal of torture – as are confession and intimidation by their absence – because the whole of the Government's activity is (falsely) discredited as torture. Echoing the negation, 'we're not punishing, we're curing,' is its reformulation, 'we're not harming, we're helping.' This tactic of moving torture out of 'torture', distancing the content from the concept, is aided by the insertion of the first without the sheath of the second into a bureaucratic order not merely semblance but fully assembled – that is, a structure capable of self-deception. Such a structure is voided of 'torturers' and filled instead with what Foucault would call 'a whole army of technicians'.[19] Foucault goes on to expand on this with 'warders,

18 Burgess, *A Clockwork Orange*, 69. Cf. the words of Foucault concerning such 'theoretical disavowal', which accord sharply with *ACO*'s insights: 'do not imagine that the sentences we judges pass are activated by a desire to punish; they are intended to correct, reclaim, 'cure'…' (*Discipline and Punish*, 10).

19 Foucault, *Discipline and Punish*, 11.

doctors, chaplains, psychiatrists, psychologists, [and] educationalists'. He also goes on to say that this mass of technicians sponsors an apparently humane 'restraint', what he poetically calls '[a] utopia of judicial reticence'.[20] Scarry, likewise, is alert to this displacement, dealing particularly with the presence in torture of 'two of civilization's institutions': the law and medicine.[21]

In *A Clockwork Orange*, the force that treats Alex very conspicuously includes medical teams – naturally for a treatment program – led by the character of Dr. Brodsky. Still, giveaways that the Government's Reclamation Treatment ought to be conceived as torture remain. Most explicitly there is the hard resonance of violence in '*Kill* the criminal reflex.' There is also the layered ominousness of the name Dr. Brodsky. The obvious Russian origin carries to the Anglophone ear a prejudice of villainy and thus malignant intent. The final syllable hints at Skinnerian behaviorism, which Burgess opposed for its inhumanity, and which bears clearer allusion in the spliced protocols of brainwashing and chemical castration. Finally, there is Alex's predicament as 'patient', traditionally 'the term used to designate the victim' of torture following Foucault.[22]

Turning to the role of imagination, this is a facet of torture silent, if not missing from, the legal-philosophical assessments but prominent in the works of fiction. Arguably, the literature sues for the expansion of torture's structure to accommodate clear recognition of the part played by imagination. Imagination is active in the creation of fictions. What can be missed in this axiomatic view is that imagination also actively shapes reality. This is acutely crucial to torture because torture is in reality a kind of performance. About this both Foucault and Scarry are emphatic,[23] which makes their relegation of imagination to structural insignificance surprising. Prerequisite to performance is conceptualisation. In order to be performed torture must first be imagined. Yet the insistence critically is that the seminal stage of torture is pain. This is not wholly incorrect. Pain is the seminal stage of the performance. But it is inaccurate. Torture does not start as action. It starts as premeditation. Instrumentality reappears here. Torture is instrumental rather than reactive violence.[24] Its goal-orientedness, regardless of what is taken as *a* or *the* goal, confesses to

20 Ibid.

21 Scarry, *The Body in Pain*, 41–42.

22 Foucault, *Discipline and Punish*, 40.

23 See: Ibid, 8, 9, 34; and Scarry, *The Body in Pain*, 10, 27, 28.

24 That there is a distinction is not a matter of unanimity. Cf., for example, Hannah Arendt, *On Violence*, (New York and London, Harcourt, 1970) and James Blair, Derek Robert and Karina Blair, *The Psychopath: Emotion and the Brain*, (Malden, MA, Blackwell, 2005).

design. Precisely the presence of a whole army of technicians in Ludovico's Technique is what betrays design in *A Clockwork Orange*. Alex is not simply subjected to torturous treatment. The character is subjected to a totalitarian behaviorist philosophy preliminary to, and surrounding, the treatment. Ludovico's Technique is the product of rigorous thought. Presumably, indeed, it is the result of what went on in the head of an otherwise unfeatured character called Ludovico.

What reveals the imaginative aspect in *American Psycho* is Patrick's own admission of it: 'On a Wednesday night another girl, who ... I *plan* to torture and film' (emphasis added);[25] and earlier, in the fortieth chapter, 'Rat', he states: 'I sit in the kitchen thinking of ways to torture girls with this animal'. To this the narrator-protagonist tacks on a disturbingly wry addendum in parentheses: '(unsurprisingly I come up with a lot)'.[26] Of all the ways the character imagines, what Patrick elects to do is insert the animal into the vagina of his victim in order that it should eat its way out of her. Noteworthy is that this gruesome act does not owe its origin to Patrick's fantasies, nor to those of the novel's author. Both anal and vaginal violation using rats is a documented torture technique. For example, it was reportedly used in Argentina by the military regime which ruled from 1976 to 1983 after deposing President Isabel Peron.[27] Crucially, this doesn't change the fact that somebody had to imagine doing such a thing before it could be done.[28] If anything, it serves to highlight it.

Alongside these overt declarations of premeditation, marking the imagination in action, there are two further suggestions of it, increasingly subtle. In both novels torture is literally screened. That is, it appears televisually within the narratives. Alex is tortured by being made to watch a series of film clips featuring various forms of violence – what one smirking technician calls '[a] real show of horrors'.[29] In accordance with Scarry's principle of ever-intensification,

25 Ellis, *American Psycho*, 326.

26 Ibid, 309.

27 For example: Robert I. Simon, *Bad Men Do What Good Men Dream: A Forensic Psychiatrist Illuminates the Darker Side of Human Behavior*, (Washington and London, American Psychiatric Press, Inc., 1996) 12–13; and, Rafael Azul and Bill Vann, 'US Documents Implicate Kissinger in Argentine Atrocities', *World Socialist Website*, accessed 2 June 2016, http://www.wsws.org/en/articles/2002/09/arg-s06.html.

28 Fuller consideration of the sadistic potential of the rat, both in the imagination and in reality, could usefully begin with a reading of Freud's "Rat Man" (*Standard Edition Volume 10: Two Case Histories "Little Hans" and the "Rat Man"*, trans. James Strachey (London: Vintage, 1975 [1909]).

29 Burgess, *A Clockwork Orange*, 76.

the five clips move along a trajectory of escalation, and themselves explicitly include torture. For example, the third focuses on:

> just a human litso [a face] ... having different nasty veshches done to it ... first a britva [a knife] cut[s] out an eye, then slice[s] down the cheek, then [goes] rip rip rip all over ... Then all the teeth [are] like wrenched out with a pair of pliers, and the creeching and the blood [are] terrific.[30]

Alex is forced to consume the product of somebody's imagination (Ludovico's again, or Dr. Brodsky's?), brought to life presumably by performers – although on this point Alex is suggestively troubled:

> This was real, very real, though if you thought about it properly you couldn't imagine lewdies actually agreeing to having all this done to them in a film, and if these films were made by the Good or the State you couldn't imagine them being allowed to take these films without like interfering with what was going on.[31]

Part of the young woman's torture in *American Psycho* is likewise being made to watch what happened to a previous victim. Following Patrick's narration: 'I've situated the body in front of the new Toshiba television set ... and appearing on the screen is the last girl I filmed ... I'm kneeling on the floor beside a corpse, eating the girl's brain.'[32] Patrick also notes: 'when she sees me, standing over her, naked, I can imagine that my virtual absence of humanity fills her with mind-bending horror.'[33] This is both apposite and cryptic. Centered in the syntactical uncertainty of the nakedness referenced – is Patrick talking about his victim's categorical nudity or his own probable nudity? – is the duplicity of two imaginations at work. His thoughts, at least in that moment, are preoccupied by the preoccupations of hers. That torture replays the fact of its imaginativeness during its performance in the fiction accords with convention in reality. Torture typically does demand in its process its own re-imagination, a point emphasised by both Scarry and Foucault. Scarry: 'Testimony given by torture victims from many different countries almost inevitably includes descriptions of being made to stare at the weapons with which they were about

30 Ibid, 78.
31 Ibid, 77.
32 Ellis, *American Psycho*, 327–328.
33 Ibid, 327.

to be hurt';[34] and Foucault: 'The first degree of torture was the sight of the instruments.'[35] Whether enacted by an authorised agent or a sadist, the potential of the imagination to pre-play the ordeal is prized in torture. In either case the torturer concedes in the moment they begin the torture in the presence of the victim that actually the torture began in the victim's absence when it was imagined by the torturer or those directing them. This covert concession is even more covertly alluded to in the novels by their very structure. Fiction demands the involvement of the imagination to be borne out. What the reader is demanded to imagine by *A Clockwork Orange* and *American Psycho* is torture, '[a] real show of horrors', and in their acquiescence the reader unwittingly acknowledges that imagination is indeed foundational to torture's structure.

This retrospective shift in the seminal stage of torture by the literature is valuable because it warns that the practice of torture is likely to be unending. This warning, coincidentally, poses a challenge to the ethos of the legal-philosophical assessments. Consciously or not, the assessments are party to the inherently cessationist aim of the current episteme's expansion of torture within the field of knowledge: fill up enough epistemological space with inimical thoughts on torture and perhaps there won't be much left for acts of it (an aim to which this chapter is admittedly also party). More than in any other element of torture, this warning resides in the monstrosity of the tortured body. Given the etymology, this is unsurprising. The English 'monster' is derived from the Latin 'monstrum', meaning 'a portent', rooted in the infinitive 'monere', 'to warn'. The tortured body, seemingly rendered as least visible as possible by the critical assessments, has long been a spectacle for literature.

Handling content millennia older than himself in *Library*, the first-century Greek mythographer Apollodorus wrote the following about the Hellenic maker of mankind:

> Prometheus fashioned humans from water and earth. He also gave them fire without Zeus' knowledge by hiding it in a fennel stalk. When Zeus discovered this, he ordered Hephaistos to nail his body to Mount Caucasus (this is a mountain in Scythia). Prometheus was nailed to it and bound for many years. Each day an eagle flew down to him and would eat the lobes of his liver, which grew back at night.[36]

34 Scarry, *The Body in Pain*, 27.

35 Foucault, *Discipline and Punish*, 40.

36 1.45 in R. Scott Smith and Stephen M. Trzaskoma, trans., *Apollodorus' Library and Hyginus' Fabulae: Two Handbooks of Greek Mythology* (Indianapolis and Cambridge: Hackett, 2007), 7.

Grappling like Apollodorus before him with a tradition, James Joyce described the underworld of Catholicism as follows in *A Portrait of the Artist as a Young Man*:

> [T]he burning ocean of hell ... will not afflict the bodies of the damned only from without but each lost soul will be a hell unto itself, the boundless fire raging in its very vitals. O, how terrible is the lot of those wretched beings! The blood seethes and boils in the veins, the brains are boiling in the skull, the heart in the breast glowing and bursting, the bowels a redhot mass of burning pulp, the tender eyes flaming like molten balls ... [37]

Although Prometheus is saved by Hercules from the torture set for him by his violently capricious father, the design in both cases is endlessness. Coded in the population of this beginning and ending by tortured bodies is the persistence of torture, beyond which is the persistence of violence. 'It happened, therefore it can happen again.'[38] This is the warning made by Primo Levi about the enormous instrumental violence of the Holocaust. Referring to the localized instrumental violence of torture, Levi's warning can be reformulated: it can be imagined, therefore it can be made to happen. One lesson manifestly available in *A Clockwork Orange*, *American Psycho* and their antecedents is that torture has lived in the imagination of mankind for centuries. Another, more latent, is that it will likely breech mankind's reality for many more. Ultimately, the literature's configuration of the seminal stage of torture as its genesis in the imagination affords the realization that if torture is to be prevented in action it would have to be vanished from imagination. But without recourse to the sort of curative totalitarian behaviorism inflicted on Alex in *A Clockwork Orange*, the latter would be impossible, binding the former to be the same. With cruelty more generally natural to the human animal,[39] imagining and then performing torture is likewise. Accordingly, people are condemned to contend continually with it.

---

37 III.694–702 in James Joyce, *A Portrait of the Artist as a Young Man*, eds. Hans Walter Gabler and Walter Hettche (New York and London: Garland, 1993 [1916]), 145.

38 Primo Levi, *The Drowned and the Saved*, trans. Raymond Rosenthal (London: Abacus, 2003 [1986]), 167.

39 Consider: Mary Midgley, *Beast and Man: The Roots of Human Nature* (Abingdon and New York: Routledge, 2002 [1979]); Ibid, *Wickedness: A philosophical essay* (London: Routledge, 1984).

Bibliography

Arendt, Hannah. *On Violence*. New York and London: Harcourt, 1970.

Azul, Rafael and Vann, Bill. 'US Documents Implicate Kissinger in Argentine Atrocities.' *World Socialist Website*. Accessed 2 June 2016. http://www.wsws.org/en/articles/2002/09/arg-s06.html.

Blair, James, Robert, Derek and Blair, Katrina. *The Psychopath: Emotion and the Brain*. Malden, MA: Blackwell, 2005.

Burgess, Anthony. *A Clockwork Orange*. London: Penguin, 2000 [1962].

Ellis, Bret Easton. *American Psycho*. New York: Picador, 1991.

Foucault, Michel. *Discipline and Punish: The Birth of the Prison*. Translated by Alan Sheridan. London: Penguin, 1991 [1975].

Freud, Sigmund. *Standard Edition Volume 10: Two Case Histories "Little Hans" and the "Rat Man"*. Translated by James Strachey. London: Vintage, 1975 [1909].

Joyce, James. *A Portrait of the Artist as a Young Man*. Edited by Hans Walter Gabler and Walter Hettche. New York and London: Garland, 1993 [1916].

Levi, Primo. *The Drowned and the Saved*. Translated by Raymond Rosenthal. London: Abacus, 2003 [1986].

Midgley, Mary. *Beast and Man: The Roots of Human Nature*. Abingdon and New York: Routledge, 2002 [1979].

Midgley, Mary. *Wickedness: A philosophical essay*. London: Routledge, 1984.

Scarry, Elaine. *The Body in Pain: The Making and Unmaking of the World*. New York and Oxford: Oxford University Press, 1987 [1985].

Simon, Robert I. *Bad Men Do What Good Men Dream: A Forensic Psychiatrist Illuminates the Darker Side of Human Behavior*. Washington and London: American Psychiatric Press, 1996.

Smith, R. Scott and Trzaskoma, Stephen M., trans. *Apollodorus' Library and Hyginus' Fabulae: Two Handbooks of Greek Mythology*. Indianapolis and Cambridge: Hackett, 2007.

United Nations. 'Convention against Torture and Other Cruel, Inhuman or Degrading Treatment or Punishment 1984.' Accessed 25 February 2016. http://www.un.org/documents/ga/res/39/a39r046.htm.

Index

www.ingramcontent.com/pod-product-compliance
Lightning Source LLC
Chambersburg PA
CBHW071346230926
51846CB00002B/460
* 9 7 8 9 0 0 4 4 0 5 7 0 7 *